NEUE SCHRIFTEN. NEW TYPEFACES.

POSITIONEN UND PERSPEKTIVEN

Herausgegeben von
Isabel Naegele, Petra Eisele, Annette Ludwig

Niggli

CONTENTS

EINLEITUNG.
INTRO.

INTERVIEWS UND SCHRIFTEN.
INTERVIEWS AND TYPEFACES.

CALL FOR TYPE.

ANHANG.
APPENDIX.

NEUE SCHRIFTEN. NEW TYPEFACES. POSITIONEN UND PERSPEKTIVEN.

Petra Eisele, Isabel Naegele, Annette Ludwig

Wir begegnen den Schriftzeichen unseres lateinischen Alphabets tagtäglich, zumeist ohne sie bewusst wahrzunehmen. Beim Lesen der Zeitung, beim Surfen im Netz – Schrift erschließt uns einen Großteil unserer Welt und transportiert Informationen nicht nur im Hier und Jetzt, sondern vermittelt über Generationen hinweg: Schriftzeichen tragen Wissen durch die Zeit. Aus dieser alltäglichen Perspektive ist Schrift im wahrsten Sinne des Wortes eine Dienerin, die uns hilft, die verschiedensten Inhalte zu kommunizieren.

Kommunikations-Designerinnen und -Designer machen es sich zur Aufgabe, die Übermittlung von Informationen mittels Schrift möglichst adäquat zu gestalten. Dies geschieht zum einen mittels Typografie, der Gestaltung durch und mit Schrift. Bei der mikrotypografischen Gestaltung stellt die Wahl einer bestimmten Schrift, ihre Zurichtung und Formatierung, das zentrale Element dar. Die Spezifika eines Schrifttyps, ihr Charakter und ihre Ausprägungen erzielen gemeinsam mit ihrer grafischen Anordnung (Makrotypografie) eine Wirkung, die über den »Transport« des Textinhaltes weit hinaus reicht. Daher resultiert die ästhetische Qualität typografischer Gestaltung aus dem stimmigen Verhältnis zwischen Form, Inhalt und Zweck.

Zum anderen geht es beim Schriftentwurf, dem sogenannten Typeface-Design, um die Gestaltung von Schrift, einer Disziplin innerhalb der Typografie, der die vorliegende Publikation gewidmet ist. Die Entwicklung von Schrift stellt die Suche nach der speziellen Form einer Schrift dar, mit der sprachliche Botschaften, aber auch nichtsprachliche Implikationen bestmöglich übermittelt werden können. Hier kann gestalterische Qualität erst dann erreicht werden, wenn die einzelnen Buchstaben – selbst diejenigen, die expressiver Eigenständigkeit geschuldet sind – so harmonisch aufeinander abgestimmt sind, dass sie in ihrer Gesamtheit als Schrift bzw. als Text wahrgenommen werden können.

Neue Software-Programme und Vertriebswege haben in den vergangenen Jahrzehnten den Entwurf von Schriften grundlegend verändert. Sie führten zu einem enormen Anstieg von Schriftproduktionen, aber auch zu einer großen Unübersichtlichkeit. Wurden in den 1970er Jahren jährlich noch wenige hundert Schriften veröffentlicht, können wir heute zwischen zehntausenden Fonts wählen — Tendenz steigend. Dass sich aktuelles Type Design dennoch keinesfalls in Beliebigkeit oder einem modischen »Hype« erschöpft, verdeutlicht diese Publikation: Aus dem reichen, internationalen Schriftkosmos präsentiert sie experimentelle Schriftentwürfe, die von einer spontanen Idee oder der Konzeption eines selbst gewählten Gestaltungsprojekts ausgehen; Display-Fonts mit wenigen Schnitten oder Leseschriften mit umfangreichen Schriftfamilien, die, oft über Jahre hinweg, mit leidenschaftlicher Ausdauer und einer unerschütterlichen Liebe zu den charakteristischen Details systematisch ausgebaut wurden. In ihrer Gesamtheit dokumentiert die Publikation »Neue Schriften. New Typefaces.« daher nicht nur die hohe Professionalität, sondern auch den ernsthaften Enthusiasmus, der Schriftgestalterinnen und -gestalter heute antreibt.

Siebzig Schriftentwürfe bieten einen Ein- und Überblick über die aktuellen Entwicklungen. Für die Ausstellung »Call for Type. Neue Schriften. New Typefaces.« im Mainzer Gutenberg-Museum wurden zwanzig Schriftentwürfe ausgewählt und durch persönliche Reflexionen der Schriftgestalterinnen und -gestalter inhaltlich ergänzt. Diese Interviews entstanden in der Vorbereitung der Ausstellung sowie im Rahmen der ausstellungsbegleitenden Veranstaltungsreihe »Talks. Gespräche über Schrift«. Hier geben Gestalterinnen und Gestalter Einblick in den Entstehungsprozess einzelner Schriften, verweisen auf Inspirationsquellen und lassen an der Faszination für Schriftentwurf teilhaben, die alle vorgestellten Fonts motiviert hat. Weitere fünfzig Schriftentwürfe wurden aus dem Pool der Einreichungen zu dem vom Gutenberg-Museum und dem Designlabor Gutenberg / Fachhochschule Mainz international ausgelobten Wettbewerb »Call for Type« ausgewählt.

In ihrer Gesamtheit erlauben es die individuellen Bewertungen und Einschätzungen formalästhetischer und inhaltlicher Aspekte, aktuelle Entwicklungen im Font Design als

Ausdruck von und Reaktion auf aktuelle technische, ästhetische und gesellschaftspolitische Entwicklungen zu fassen.

ZWISCHEN FORMEXPERIMENT UND REGELHAFTIGKEIT

Die zum Teil sehr persönlichen Ausführungen der Schriftgestalterinnen und -gestalter machen deutlich, dass die Genese einer Display-Schrift oft der Einstieg in ein spezielles Gestaltungsprojekt ist, um dessen »Sound« zu unterstützen oder um ihm eine individuelle »Handschrift« zu verleihen. Die Schrift wird hier Ausdruck von Identität. Oft wird die Entwicklung einer Schrift als eine Art Spiel begriffen, in dem freie Formexperimente gegen gestalterische Traditionen und Konventionen gesetzt werden. Dabei bildet die Glyphe den Ausgangspunkt für selbst definierte Spielregeln bzw. selbst auferlegte Formzwänge, sogenannte »Constraints«. Sie definieren eine individuelle Ausformung der Letter wie bei der HM Tilm → S. 114, für die eine zeitliche Begrenzung auf 24 Stunden Bearbeitungszeit den Rahmen für die Formfindung vorgab. Nicht selten führen auch Materialexperimente oder Zufallsfunde zu eigenwilligen Display-Schriften, wie bei der Happypeppy → S. 106, die aus einem einfachen Geschenkband geformt wurde.

Der Widerstand gegen normative Tendenzen und die Verneinung einer definierten Form motivierte die Gestaltung der Lÿno → S. 140. Bei der MeM → S. 158 rückte die Lesbarkeit in den Hintergrund, so dass die Grenzen zwischen Grafik-Design und Schriftgestaltung verschwimmen. Zahlreiche Alternativzeichen und die extremen Stile der einzelnen Glyphen erzeugen eine lebendige Spannung aus geometrischer Anmutung und verspielten Formen, die eine kontrastreiche Mischung von ausdrucksstarken und feinteiligen Elementen ergibt, in der jeder Buchstabe zu einer eigenständigen Grafik avanciert.

Die GT Haptik → S. 96 dagegen wurde inspiriert durch die Ergebnisse von Untersuchungen über den Tastsinn. Daher sind ihre Versalien und Zahlen auf die einfachste Form kondensiert. Die hieraus resultierenden Buchstabenformen und monolinearen Strichstärken dominieren den eigenwilligen Entwurf. Erst in einem zweiten Schritt wurden Strichstärkenkontraste und optische Korrekturen durchgeführt.

Die Faszination für technische Konstruktionen kann ebenfalls ein Leitmotiv für die Entstehung von Schriftzeichen sein. So orientiert sich die AB Eiffel → S. 20 unverkennbar an der charakteristischen Stahlkonstruktion des Pariser Eiffelturms, also dem Quadrat, den Doppelquadraten und den Verstrebungen in verschiedenen Winkeln.

NACH DER GESCHICHTE

Nach wie vor werden historische Vorbilder analysiert und neu interpretiert – als Hommage an die großen Meister, aber auch im Sinne der Schulung des eigenen Qualitätsanspruches und der Bewusstmachung der eigenen Tradition.

So widmet sich die Capitolium News → S. 56, die 2006 entworfen wurde, einem in der Tat klassischen Thema, nämlich der römischen Schrifttradition: Ursprünglich aus Anlass des 2000-jährigen Jubiläums der römisch-katholischen Kirche für ein Informations- und Leitsystem durch die »ewige Stadt« für Pilger und Touristen entwickelt, reflektiert diese Schrift auch die zweitausend Jahre alte Tradition öffentlicher Beschriftung in der italienischen Metropole. Sie wurde mit den Ansprüchen an eine Zeitungsschrift verbunden, was in dem Zusatz »News« sinnfällig gekennzeichnet ist.

Dass die Gestaltung einer Leseschrift ernsthaften Enthusiasmus oft über Jahre oder gar Jahrzehnte erfordert, verdeutlichen die Roletta → S. 176 sowie die Axia → S. 40. Sie zeichnen sich durch eine große Auswahl an OpenType-Features sowie zahlreiche Schriftstile aus, die von Grund auf entwickelt und auf gute Lesbarkeit abgestimmt wurden. Diese umfangreichen Schriftfamilien verbinden Funktionalität mit einem spielerischen Duktus.

Das Thema »Futuretro«, bei dem sich Designerinnen und Designer produktiv mit der Vergangenheit auseinandersetzen, indem sie sich diese gestalterisch neu aneignen, manifestiert sich auch im Type Design. Aus dem luziden Potential einer einfachen Schriftschablone aus dem Jahr 1876 beispielsweise entstand mit der PDU → S. 166 eine modulare Schriftenfamilie. Die Maison bzw. Maison Neue → S. 148 mit ihren 12 Gewichten dagegen orientiert sich an den Prinzipien alter Groteskschriften (u.a. der Akzidenz Grotesk), um eine zeitgemäße Grotesk entstehen zu lassen.

Die Wiederentdeckung historischer Schriften, insbesondere aus den 1920er und 1930er Jahren, deren individuelle Gestaltungsmerkmale häufig der Digitalisierung und dem Zeitgeschmack zum Opfer fielen, findet ebenfalls sichtbaren Niederschlag. Sie dienen in jüngster Zeit verstärkt als »Ausgangsmaterial« für gestalterische Neuinterpretationen wie bei der Brandon Grotesque → S. 48, die durch ihre geringe x-Höhe eine besondere Eleganz erhält. Dass Schrift sogar deutsch-deutsche Vergangenheit fassen kann, verdeutlicht die Karbid, die Laden- und Fassadenbeschriftungen aus Ost-Berlin rezipierte, die heute längst verschwunden sind. Die 2011 re-designte und mit ausgebauten Schriftfamilien erhältliche Karbid Pro → S. 122 verarbeitete das Verlorene neu, ohne sich in einem nostalgisch-vergangenheitsverklärenden Retro-Design zu verlieren.

Schriftmuster der 1970er Jahre, wie die Lettera-Bücher des Schweizer Niggli-Verlags, inspirierten die Grow → S. 88 mit technisch vielschichtigen Zeichen und einer nahezu unüberschaubaren Vielzahl an Kombinationen, wie sie heute nur in Zusammenarbeit mit sogenannten Type-Engineers verwirklicht werden können.

Inspirationsquellen sind aber auch freie assoziative Zugänge zur Kunst- oder Designgeschichte, wie sie Karl Nawrot für die Corporate Identity der Stiftung Bauhaus Dessau entwickelte. Zu den vier Schriften seiner Dess → S. 64 (Josa, Breu, Mona und Pauk) ließ er sich aus dem gestalterisch-künstlerischen Œuvre der Bauhaus-Meister Josef Albers, Marcel Breuer, László Moholy-Nagy und Paul Klee anregen.

HYBRIDE IDENTITÄTEN

Oft geht es im Type Design um eigenmotivierte Projekte, die eine bestimmte gestalterische Konzeption vorantreiben und umsetzen, beispielsweise indem verschiedene

Schrift-Identitäten wie bei einem genetischen Experiment miteinander gekreuzt und vereinigt werden.

Für die Fraktendon Pro → S. 70 wurden zwei vollkommen unterschiedliche Schriften, eine Fraktur und die Clarendon, miteinander »vermählt« und als sogenannter »Bastard« ausgearbeitet. Die Karloff → S. 130 verbindet eine hochkontrastierende, von Bodoni und Didot inspirierte Schrift, mit der monströsen »Italian«. Der Unterschied zwischen der attraktiven und der irritierenden Schrift wird von einem Design-Parameter bestimmt, nämlich dem Kontrast zwischen fett und dünn. Nachdem im ersten Schritt zwei völlig gegensätzliche Versionen entworfen wurden, wagten sich die Type Designer an ein genetisches Experiment mit der Schönen und dem Biest, einer Interpolation zwischen zwei Extremen, und schufen so eine Schrift mit hybrider Identität.

KULTURELLE IDENTITÄTEN

Die grenzenlose Verfügbarkeit gestalterischer Traditionen und Schriften führte auch dazu, dass vermeintlich klar definierte nationale Identitäten neu verhandelt wurden. Der besonders prägnante und international erfolgreiche »Swiss Style« beispielsweise ist von den Nachgeborenen so stark verinnerlicht worden, dass inzwischen in England, so die Einschätzung von André Baldinger, mehr »schweizerisches« Grafik-Design entsteht als in der Schweiz.

Seit einigen Jahren schlägt Typeface-Design verstärkt Brücken zwischen den großen Gestaltungstraditionen des lateinischen Schriftsystems und anderen Schriftsystemen, vornehmlich des osteuropäischen (Adelle → S. 30) und arabischen (Gebran2005 → S. 78) Raums. Damit vermittelt Type Design ganz real zwischen den unterschiedlichen Kulturen und trägt zu Verständnis, Ausgleich und Toleranz bei, indem es das jeweils Eigene selbstverständlich zum Teil des anderen werden lässt, ohne es seiner Identität zu berauben.

Nicht nur zahlreiche Schriften, sondern auch viele Lebensläufe von Schriftgestaltern changieren zwischen den Kulturen und entwickeln eine hohe gesellschaftspolitische Relevanz. Der gebürtige Tscheche Peter Bil'ak etwa, der an der holländischen Jan van Eyck Akademie studierte und heute als Grafiker und Schriftgestalter in den Niederlanden tätig ist, ist unter anderem Mitbegründer der »Indian Type Foundry« in Ahmedabad, im nördlichen Indien, die unter anderem seine Fedra Hindi (2007–2009 Peter Bil'ak und SN Rajpurohit) vertreibt und zu dem rasant wachsenden Interesse und zu der Professionalisierung des indischen Type Design entschieden beiträgt.

Auch aus einer anderen Perspektive wirkt Type Design politisch und im wahrsten Sinne des Wortes gesellschaftsbildend: Die libanesische Schriftgestalterin Nadine Chahine betont, dass mit Hilfe gut gestalteter und damit leichter lesbarer Schriften auch dem Problem des Analphabetismus und seinen Auswirkungen entgegengewirkt werden kann.

Wir wünschen uns, dass die Faszination am Type Design, die uns die Anregung und Motivation für die Ausstellung »Call for Type« und diese Publikation gegeben hat, für Schriftgestalterinnen und Schriftgestalter, aber auch für allgemein an Typografie Interessierte, eine Inspirationsquelle darstellen und die Sensibilität für Schriftgestaltung und ihre Protagonisten erhöhen kann. Anlässlich unseres »Gesprächs über Schrift« hat denn auch Jost Hochuli treffend bemerkt: »Nur wer Freude am Lesen und an Büchern hat, bekommt ein Gefühl für Buch und Schrift. Wenn beim Lesen ganz bewusst auch auf die Wirkung der Typografie geachtet wird, lernt man viel.«

NEUE SCHRIFTEN. NEW TYPEFACES. POSITIONS AND PERSPECTIVES.

Petra Eisele, Isabel Naegele, Annette Ludwig

Not a day goes by where we are not confronted with our Latin characters, usually without consciously registering them. Whether we read the paper or go online — the written word opens the world for us and conveys information, not just here and now, but across generations: written characters carry knowledge across time. Seen from that angle, writing is, in the word's truest sense, a servant which helps us to communicate contents embracing a vast range of topics.

Communication designers make it their job to organise this conveyance of information via type as efficiently as possible. This happens, on the one hand, via typography — the design with and through type. In micro-typographic design, the choice of a type, its kerning and formating, are the central elements. The particulars of a type, its character and distinctness, together with the graphic arrangement (macro-typography) have an impact far beyond the »transport« of the contents of a text. Thus, the aesthetic quality of typographic design is achieved by a harmonious relationship between form, contents and purpose.

Typeface design, however, deals with the design of type itself, a very specialized discipline within typography, to which we dedicate this publication. The development of a typeface represents the process of searching for type of a special form with which linguistic messages, but also non-linguistic implications can be transmitted in the best way possible. Here, a good design quality can only be achieved when all single characters — even those resulting from expressive individuality — are so much in harmony with each other that they will be perceived in their entirety as a typeface or a text, respectively.

In recent years, new software programmes and distribution channels have radically changed type design. This resulted in a vast increase of type production, but also in a lot of confusion. While in the 1970s just a few hundred new typefaces were published, today we can choose from tens of thousands of fonts — with upward tendency. This publication, however, shows that current type design is not subject to randomness or a fashionable »Hype«: from the rich international cosmos of typefaces it presents experimental type designs, based either on a spontaneous idea or the concept of an individually chosen design project; there are display fonts with a small range of styles or typefaces for books with large type families, which, often over a number of years, were developed systematically with passion, endurance and an unswerving attention to characteristic detail. Therefore this publication »Neue Schriften. New Typefaces.« not only documents a high professionalism, but also a great and passionate enthusiasm as the driving force of today's type designers.

Seventy different type designs present an overview of current developments. For the exhibition »Call for Type. Neue Schriften. New Typefaces.« at the Gutenberg-Museum in Mainz twenty typefaces were chosen and expanded on in very personal reflections by their designers. These interviews came about during the preparation for the exhibition and the accompanying series of lectures »Talks. Gespräche über Schrift«. Here the designers give us a glimpse into the genesis of some of their typefaces, they talk about their sources of inspiration and share with us their fascination for type design. A further fifty typefaces were chosen from the pool of entries submitted for the international competition »Call for Type«, staged by the Gutenberg-Museum and the Design Laboratory Gutenberg / Fachhochschule Mainz.

As a summary, the individual assessments and opinions about aspects of formal-aesthetic nature and content may be used for the formulation of theses and help to see current developments in font design as an expression of and a reaction to technical, aesthetic and socio-political developments. This not only strengthens our awareness of the hitherto neglected historic, political and social aspects of type design, but demonstrates that type design is also part of the current discourse about the contentual and political dimensions of design as such.

FORM — BETWEEN EXPERIMENT AND REGULARITY

The — sometimes highly personal — deliberations of the various type designers very clearly indicate that the genesis of a display font often coincides with the beginning of a special design project in order to emphasize its »sound« or to give it an individual »thumbprint«. Here, type becomes an expression of individuality. Frequently the development of a typeface is considered a kind of game, in which free form experiments are contrasted with design traditions and conventions. Here, the glyph is the starting point for the self-defined rules of the designer, the so-called »constraints«. They define the individual shaping of a letter, as in the HM Tilm → S. 114, for which the time for finding the form was limited to 24 hours. Sometimes experiments with materials or random discoveries will lead to very unconventional display fonts, as in the Happypeppy → S. 106, which was created from a simple gift ribbon.

The design of the Lÿno → S. 140 was motivated by the rejection of normative tendencies and the negation of a defined form. With MeM → S. 158 legibility was pushed into the background, resulting in a fluid borderline between graphic design and type design.

Numerous alternative characters and the extreme styles of single glyphs create a vivid tension between geometric impressions and playful forms, resulting in a mix of expressive and delicate elements where each letter turns into an individual piece of graphic art.

GT Haptik → S. 96, on the other hand, was inspired by the results of a study about the tactile sense. Thus, its capital letters and numbers are reduced to their simplest forms. The resulting letter forms and mono-linear stroke widths dominate this idiosyncratic design. Optical corrections and contrasts in stroke widths were only added in a second step.

Fascination for technical construction can also serve as an inspiration for a new typeface. Thus, the AB Eiffel → S. 20 unmistakably echoes the characteristic steel construction of the Eiffel Tower in Paris with its squares, the double squares and the differently-angled struts.

FROM HISTORY

To this day historic models are being analyzed and newly interpreted — as an homage to the great masters, but also in order to sharpen one's own demand for quality and consciousness of tradition.

Thus, the Capitolium News → S. 56, created in 2006, did indeed take up a classic theme, i.e. the Roman tradition of writing. Originally developed for the 2000 year jubilee of the Roman-Catholic church for a guiding and information system for pilgrims and tourists through the »eternal city«, it also reflects the 2000 year-old tradition of public lettering within the Italian capital. The requirements of a newspaper also had to be considered, as reflected in the addition of »News« in the name.

The enthusiasm for developing a typeface over years or even decades is demonstrated by the Roletta → S. 176 and the Axia → S. 40. They stand out for their large number of OpenType features and their many styles, which were developed from scratch and fine-tuned for good legibility. These large type families combine functionality with a playful style.

The subject of Futuretro, where designers are dealing with the past in a playful way by giving them new design forms, is also manifest in type design. From the lucid potential of a simple letter stencil from 1876, for instance, we got the modular type family PDU → S. 166. The Maison or Maison Neue → S. 148 resp. with its 12 weights, on the other hand, echoes the principles of old grotesques (the Akzidenz Grotesk, among others) in order to arrive at a contemporary grotesque.

The rediscovery of historic typefaces, in particular those from the Nineteen-Twenties and Thirties, whose individual design characteristics often fell victim to digitalization and prevailing taste, is also evident. In recent years this has frequently been the starting point for new design interpretations, as in the Brandon Grotesque → S. 48, which has a particular elegance through the low x-height. The Karbid shows that even the youngest German history had its special typefaces, for this was used for shopfronts and display in East Berlin, where it has long since disappeared. The Karbid Pro → S. 122, re-designed in 2011 and developed into a large type family, is using the old type without getting lost in nostalgic retrodesign.

Type patterns of the Seventies, like the Lettera books by the Swiss publisher Niggli, inspired the Grow → S. 88, with its technically complex characters and a multitude of combinations of a kind which can only be achieved in cooperation with type engineers.

However, the free associative access to art or design history may also serve as a source for inspiration, as has been shown by Paul Nawrot, who developed the Corporate Identity of the

Bauhaus Dessau Foundation. For the four typefaces of his Dess → S. 64 (Josa, Breu, Mona and Pauk) he got his inspiration from the artistic œuvre of the Bauhaus Masters Josef Albers, Marcel Breuer, László Moholy-Nagy and Paul Klee.

HYBRID IDENTITIES

In type design, self-motivated projects play a large part, they often push a certain design concept forward and implement it, for instance, by mixing or fusioning different type identities, similar to a genetic experiment.

For the Fraktendon Pro → S. 70, two completely different typefaces, a Fraktur and the Clarendon, were »married« and further developed to create a so-called »Bastard«. The Karloff → S. 130 unites a high-contrast type inspired by Bodoni and Didot with the monstrous »Italian«. The difference between the attractive and the irritating typeface is caused by one design parameter, namely the contrast between fat and thin. After – as a first step – designing two completely opposite versions, the type designers chanced a genetic experiment with Beauty and the Beast, an interpolation between two extremes, and thus created a typeface with a hybrid identity.

CULTURAL IDENTITIES

The almost unlimited availability of design traditions and typefaces also made it necessary to reconsider national identities which hitherto were thought of as being clearly defined. The particularly succinct and internationally successful »Swiss Style«, for instance, has been internalized by the younger generations to such an extent, that by now, according to André Baldinger, more »Swiss« graphic design is being created in England than in Switzerland.

For some years now typeface design builds bridges between the great design traditions of the Latin character system and other systems, particularly in the East European (Adelle → S. 30) and Arabic (Gebran2005 → S. 78) countries. Not only typefaces, but many life careers of type designers alternate between cultures and thereby develop a strong socio-political relevance. Czech-born Peter Bil'ak, for instance, who studied at the Dutch Jan van Eyck University and nowadays works as graphic and type designer in the Netherlands, is also co-founder of the »Indian Type Foundry« of Ahmedabad in Northern India, which, among others, is selling his Fedra Hindi (2007–2009, Peter Bil'ak and SN Rajpurohit), contributing to the fast-growing interest and professionalization of type design.

Type design, however, is not only political, it is also educational: the Lebanese type designer Nadine Chahine stresses that texts which are well designed and easier to read can help to overcome the problem of illiteracy and its negative effects on society.

We hope that the fascination for type design, which has inspired us to stage the exhibition »Call for Type« and compile this publication, will be a source of inspiration not only for type designers, but for everybody interested in typography, and increase the general awareness for type design and its protagonists. As Jost Hochuli so aptly remarked during our lectures »Talks. Gespräche über Schrift«: »Only people who love books and enjoy reading develop a feeling for books and type. If, while reading, you make a conscious effort to look at the effect of the typography, you can learn a lot.«

INTERVIEWS UND SCHRIFTEN.
INTERVIEWS AND TYPEFACES.

Gustave Eiffel

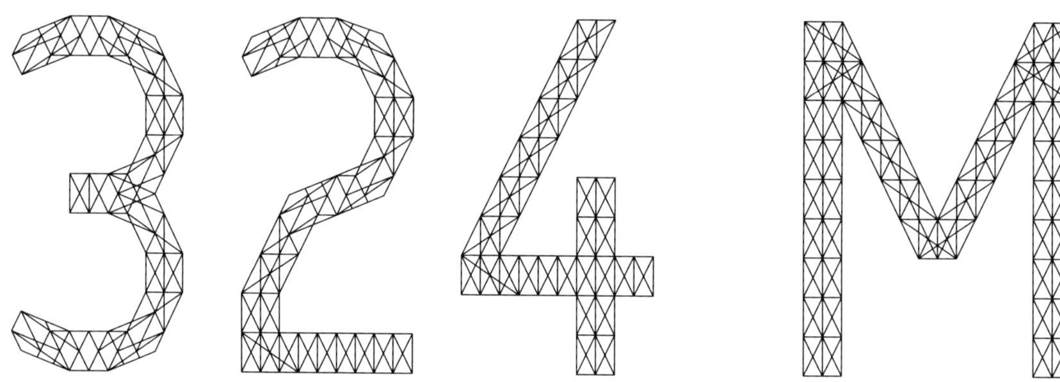

«Le plus haut bâtiment du monde»

1887 à 1889

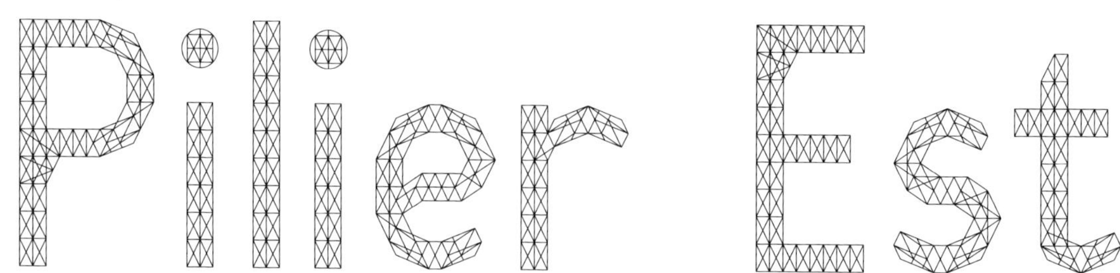

Charles Léon Stephen Sauvestre

AB Eiffel-Bold
AB Eiffel-Niveau2

AB Eiffel-Regular
AB Eiffel-Bold

AB Eiffel-Niveau2
AB Eiffel-Bold

Hohe Türme haben nicht nur aufgrund der biblischen Vorlage des Turmbaus zu Babel einen kulturellen Hintergrund, sondern gelten auch als Sinnbild für die Überwindung der Schwerkraft, als Zeichen der Herrschaft über den Raum und damit auch oft über die Menschen im Umkreis. In diesem Kontext ist der ursprüngliche Widerstand gegen den Eiffelturm als ein besonders herausragendes Beispiel der beherrschenden Macht von technischen Türmen zu sehen.

AB Eiffel-Bold – 8pt

Andererseits erfüllte Eiffel mit dem Bau seines Turms einen Menschheitstraum, nachdem rund 100 Jahre zuvor von Montgolfière bereits der Traum vom Fliegen verwirklicht worden war.

AB Eiffel-Regular – 14pt

Der Eiffelturm hatte über die architektonische Leistung hinaus eine starke Bedeutung für das französische Nationalbewusstsein.

AB Eiffel-Bold – 30pt

MONUMENTALES EINGANGSPORTAL UND AUSSICHTSTURM

AB Eiffel-Niveau2 – 50pt

Schrift. Typeface. AB Eiffel
Gestalter. Designer. André Baldinger
Label. Foundry. abtypefoundry.com
Jahr. Year. 2005–2009

Auslöser für die Entwicklung der Schrift war ein Gestaltungswettbewerb für eine neue Signaletik des Eiffelturms in Paris, zu dem Intégral Ruedi Baur Paris eingeladen war.

Die Schrift sollte für zwei sehr unterschiedliche Einsatzbereiche konzipiert werden. Eine große Display-Variante für die Beschriftung und eine Text-Variante für die Drucksachen. Während der Recherchen wurde schnell klar, dass es nicht die Form des Turmes, sondern dessen Strukturelemente sind, welche einen interessanten Ausgangspunkt für die Schriftgestaltung bilden. Charakteristisch für die Struktur sind das Quadrat, das Doppelquadrat und eine Reihe von Verstrebungen mit verschiedenen Winkeln. Diese geometrische Formensprache war das Ausgangsmaterial für den Konstruktions- und Entwurfsprozess.

The trigger for the development of this font was a design competition for a new signage of the Eiffel Tower in Paris, to which Intégral Ruedi Baur was invited. The typeface was required to serve two very different purposes: a large display variant for signs, and a text variant for printed matter.

During research it quickly became evident that it is less the shape of the tower as such, but rather its structure elements which offered an interesting starting point for the type design. The characteristics of the structure are the square, the double square and a series of struts at different angles. This geometric form language was the starting point for the construction and design process.

André Baldinger[1] (*1963) studierte bei H.R. Bosshard[2] in Zürich und absolvierte ein NDS in Schriftentwicklung/-gestaltung am Atelier National de Recherche Typographique, Paris. 1995 Gründung eines eigenen Büros, 2008 Büro für Gestaltung »Baldinger•Vu-Huu«, zusammen mit Toan Vu-Huu.[3] Baldinger ist Mitglied der Alliance Graphique Internationale, hat zahlreiche Schriften entworfen sowie visuelle Identitäten, Plakat-, Buch- und Ausstellungsgestaltungen und Szenographien. Er lehrt Schriftgestaltung unter anderem an der ZHdK[4] sowie an der EnsAD[5] Paris und ist Co-Direktor des EnsADLab Type. Zu seinen Schriften gehören die AB Baldinger Pro, die AB BDot & AB BLine-Familie, die _AB Eiffel_ sowie die AB Newut.

Stellen Sie sich bitte kurz vor.

ab: In Zürich geboren, aufgewachsen in Flims und am Greifensee. Mein Vater war Zürcher, meine Mutter ist aus Leipzig. Mit 19 Jahren zog ich nach Zürich um. Mein erstes Studium zum Elektroingenieur habe ich nach drei Semestern abgebrochen, um bei Robert Krügel, dem ehemaligen künstlerischen Direktor der Schweizer Banknoten, ein Grafikpraktikum zu machen. Später arbeitete ich mit Richard Feurer, Hannes Wettstein[6] und der neu gegründeten Design-Firma Eclat[7] zusammen. Während dieser Zeit wurde die Typografie zum bevorzugten Element in meiner Arbeit. Ich vertiefte mein Wissen im Nachdiplomstudium zum Typographischen Gestalter bei Hans Rudolf Bosshard – eine wichtige und stimulierende Begegnung. Danach hatte ich den Wunsch, Typografie nicht nur anzuwenden, sondern Schrift auch selbst zu gestalten. 1992 präsentierte ich mein Dossier am ANCT (Atelier national de création Typographique) in Paris und zog zum Nachdiplomstudium in Schriftgestaltung nach Paris. Ich begegnete Hans-Jürg Hunziker,[8] Albert Boton[9] und Adrian Frutiger,[10] verliebte mich in die Stadt und blieb.

Ich erhielt meine erste Professur an der Écal in Lausanne. Seither nimmt Grafik-Design, Schriftgestaltung und parallel dazu die Lehre einen großen Teil meiner Zeit ein. Seit mehreren Jahren erfolgt auch eine zunehmende Auseinandersetzung mit Schriftsystemen des asiatischen Sprachraumes, mit wiederholten Reisen und Aufenthalten in Japan, China und Taiwan.

Was finden Sie an Type Design interessant und aus welchem Grund haben Sie sich für diesen gestalterischen Bereich entschieden?

Typografie interessiert mich, weil ich Sprache und Inhalten visuell, konzeptuell und strukturell Ausdruck geben kann. Dabei ist die Schrift

1 andrebaldinger.com baldingervuhuu.com abtypefoundry.com

2 Hans Rudolf Bosshard (*1929) ist Schriftsetzer, Künstler und Gestalter sowie Herausgeber verschiedenster Druckerzeugnisse, zuletzt freier Buchgestalter beim Niggli-Verlag.

3 Toan Vu-Huu (*1973) wurde in Deutschland geboren, hat in Darmstadt studiert und lebt heute als Grafik-Designer in Paris.

4 Die Züricher Hochschule der Künste (ZHdK) entstand 2007 aus der Fusion der Hochschule für Gestaltung und Kunst (HGKZ) und der Hochschule für Musik und Theater (HMT). Das Museum für Gestaltung Zürich ist der ZHdK angegliedert (→ S. 90).

5 Die EnsAD, École nationale supérieure des Arts Décoratifs de Paris, wurde 1766 gegründet.

6 Hannes Wettstein (1958–2008) war ein Schweizer Industrie-Designer, der u.a. für Bosch und Lamy tätig war.

7 Die 1988 von Daniel Zentner gegründete Firma Eclat ist eine Schweizer Agentur für Markenberatung und Branding. Zu ihren Kunden gehören SSB und Swisscom (→ S. 98).

8 Hans-Jürg Hunziker (*1938) ist Schweizer Schriftgestalter. Er arbeitete u.a. mit Adrian Frutiger in Paris und entwarf die Siemens Corporate Schriftfamilie.

9 Der Grafiker Albert Boton (*1932) arbeitete als Typograf und Gestalter u.a. zusammen mit Adrian Frutiger bei Deberny & Peignot.

10 Adrian Frutiger (*1928), Schriftgestalter, der die »Schweizer Typografie« maßgeblich beeinflusste. Zu seinen bekanntesten Schriften zählen die Frutiger, die Univers, die OCR-B und die Avenir.

mein Hauptakteur. Diese kann ich noch präziser einsetzen, wenn ich sie maßgeschneidert selber gestalten kann. Schriftgestaltung ist deshalb integraler Bestandteil meiner Gestaltungspraxis. Ich denke auch, dass sie nicht nur eine exzellente Schule des Sehens und der optischen Gesetzmäßigkeiten ist, sondern auch konzeptuelle Arbeit und Zusammenhänge bewusster macht. Vom Detail zum Ganzen und zurück.

Gibt es Ihrer Meinung nach typografische Trends in der Gestaltung von Schriften sowie in deren Anwendung?

Es gibt immer Trends und Strömungen, die an die Zeit, die sozialen, politischen, kulturellen und technischen Gegebenheiten gebunden sind. Die Gegenwart ist das Resultat unserer Geschichte.

Internet und die einfache Zugänglichkeit von Information verkürzt die Lebensdauer dieser Trends zunehmend. Schrift scheint mir aber etwas unabhängiger zu sein als deren Anwendung. Im Anwendungsbereich sehe ich erschreckend viele Beispiele, die formell wohl bestehen und durchaus Qualitäten haben, aber total austauschbar sind.

Gute innovative Lösungen sind das Resultat einer intensiven Auseinandersetzung mit der Thematik und ihrem Kontext. Oberflächlichkeit ist dabei ausgeschlossen. Kultureller Hintergrund fließt immer mit in die Gestaltung ein. Die Welt wird aber immer kleiner, internationaler und dadurch leider auch oft flacher. Früher war kulturbedingte Ästhetik klarer geografisch lokal verankert. Heute kommt meiner Meinung nach aus England »schweizerischeres« Grafik-Design als aus der Schweiz. Ich selber denke, jeder Gestalter sollte seine authentische Arbeit machen, mit seinem persönlichen kulturellen Hintergrund.

Denken Sie, dass die Qualität von Schriftentwürfen durch die heute vereinfachte Zugänglichkeit zum Schriftentwurf leidet oder davon profitiert?

Der vereinfachte Zugang zur Schriftgestaltung via Fachliteratur, spezialisierten Studiengängen und immer besser werdenden Werkzeugen ist sicher positiv. Früher war dieser Gestaltungsbereich nur einer recht kleinen Gruppe zugänglich. Die Produktion des Schriftschaffens ist in den letzten zwanzig Jahren regelrecht explodiert, damit auch die Auseinandersetzung und das Bewusstsein zu Fragen rund um Schrift.

Schriftgestalter der klassischen Garde zeichneten ihre Entwürfe und gaben den technischen Teil dann ab an die Spezialisten in den Ateliers der Schrifthäuser. Heute ist dies immer weniger der Fall. Ein großer Teil der Entwerfer macht fast alles, von A bis Z. Dabei sind gerade in den letzten Jahren der technische Aufwand und das damit verbundene Wissen mit Unicode und OpenType wesentlich kniffliger und aufwändiger geworden. Auch die Bearbeitung des Delta-Hintings für eine möglichst gute Wiedergabe am Bildschirm braucht, wenn man es seriös machen will, einiges an Know-how und Zeit. Dagegen kann man heute nicht nur autonom entwerfen und realisieren, auch die Vertriebskanäle sind via Web diverser und einfacher.

Ob dabei jedoch die Qualität proportional zu der Summe der jedes Jahr realisierten Schriftentwürfe mithält, da bin ich mir nicht sicher. Aber noch nie standen so viele qualitativ hochwertige Schriften wie heute zur Verfügung.

Woher kam die Idee für die Eiffel? Und wie haben Sie die Regeln für Ihren Entwurf hergeleitet?

Eine maßgeschneiderte Schrift integriert ein Maximum an kontextueller Überlegung, ihren Einsatzbereich, das Umfeld, ihren Gebrauch und ihre Thematik.

Es war für mich klar, dass die Schrift die Einmaligkeit des Eiffelturms einfangen sollte und so diese Ikone in unverwechselbarer Art und Weise begleiten kann. Die Display-Variante durfte deshalb expressiver sein als die Text-Variante.

Die Idee kam beim Betrachten und der Analyse der Struktur des Eiffelturms. Gustave Eiffel und seine Ingenieure hatten eine Lösung gefunden, die mit einem Minimum an Material ein Maximum an Stabilität erzielt und es erst so möglich wurde, die für die damalige Zeit revolutionäre Höhe von 313 m zu erreichen.

Im Wesentlichen leiteten mich zwei Hauptparameter: Zum einen die Typologie einer leichten geometrischen Architektur, zum anderen spezifische geometrische Winkel, die sich unter anderem in dieser Struktur wiederfinden.

Dieses geometrische Formrepertoire war mein Konstruktions- und Entwurfsmaterial. Es hat mir erlaubt, der Display-Variante mit einer rein zweidimensionalen Struktur eine dreidimensionale Sensation zu geben, welche der des Eiffelturmes nahekommt.

Die Notwendigkeit, neben der Display-Variante für große Anwendungen auch eine solide (flächenfüllende) Textvariante in zwei Schriftgraden zu erarbeiten und die nach oben zunehmend sich verjüngende Struktur mit den drei Plattform-Etagen des Eiffelturmes brachten mich auf die Idee, die Display-Variante zusätzlich durch zwei Schriftstärken zu ergänzen.

So hat diese Schriftfamilie fünf Schriftschnitte. In der soliden Textvariante Regular und Bold. In der Display-Variante Niveau1, Niveau2 und Niveau3. Abgeleitet ergibt sich folgendes Schema von unten nach oben: Ebenerdig für den Mengensatz die Textvarianten AB Eiffel-Regular und -Bold. Folgen die Plattform-Etagen. Erste Plattform-Etage: AB Eiffel-Niveau1, zweite Plattform-Etage: AB Eiffel-Niveau2, dritte Plattform-Etage: AB Eiffel-Niveau3.

Welche Probleme ergeben sich bei einer derart konstruierten Schrift wie der Eiffel und wie lösen Sie diese?

Es gab formelle, optische, ästhetische und technische Probleme. Die formellen und ästhetischen Probleme waren vor allem für die Display-Varianten zu meistern. Das definierte Gestaltungsprinzip lässt weniger Spielraum als eine freie Strichführung. Es gibt kaum Modelle, an denen ich mich hätte orientieren können. Viel Zeit ging in die ideale Formfindung.

Später war die Digitalisierung eine enorme technische Herausforderung. Mit Strichstärken von fünf Einheiten (Niveau1), zwei Einheiten (Niveau2) und einer Einheit (Niveau3) fand ich mich im Grenzbereich der technischen Limitationen, die mich lange Zeit beschäftigte. Die Feinheit der Strichstärken wurde im Zusammenhang mit der Positionierung innerhalb der fixen Einheiten der Software zum Problem.
Für die Solidvarianten betrafen die Probleme vor allem optische, ästhetische Faktoren und die Kohärenz. Die einmal ausgefüllte Struktur der Display-Variante musste in einer Gratwanderung so nahe wie möglich an der Display-Variante bleibend überzeichnet werden, ohne fleckig zu erscheinen.

Dies erforderte knifflige optische Korrekturen, um die stilistischen familienbildenden Elemente beizubehalten. An den Solid-Varianten hat Anton Studer[11] mitgearbeitet und so waren wir beide im Sommer 2009 ganz schön am Schwitzen.

Wurde Ihr Entwurf bei der Beschilderung des Eiffelturms realisiert?

Der Einsatz der Schrift war für die Signaletik des Eiffelturmes vorgesehen. Die damaligen Rahmenbedingungen waren aber politisch und strategisch dermaßen komplex, dass das Projekt immer wieder verzögert wurde und schlussendlich nach politischen Wechseln nicht mehr zur Realisierung kam. Dafür wird sie dieses Jahr für ein großes Signaletikprojekt von meinem australischen Kollegen David Pidgeon[12] in Melbourne verwendet.

Ist es Ihnen generell wichtig, dass sich eine Schrift aus ihrem Inhalt ableitet oder dass sie einen Mehrwert hat, der im Zweifelsfall über ästhetischen Aspekten steht?

Ja, natürlich versuche ich jedes Mal, auch die ästhetische Seite maximal einzubinden. Schönheit und Ausgeglichenheit sind aber nicht immer die Maxime. Sie kann unter gewissen Aspekten schlichtweg langweilig oder unpassend sein. Kontext und Einsatz spielen eine wichtige Rolle.

Ein anderer Parameter ist Kohärenz. Eine Gestaltungslösung, welche nicht in hohem Maß kohärent ist, hat gleich »schlechtere Karten«. Bei Schrift kommt diesem Parameter ein besonderer Stellenwert zu.

Die meisten der von mir bisher entworfenen Schriften sind aus dem Bedürfnis entstanden, Lösungen für fehlende oder in meinen Augen unbefriedigende bestehende Schriften vorzuschlagen. In der Regel waren dies spezifische Parameter, die mich interessierten. Diese können

11 Anton Studer (*1983) ist Schweizer Grafiker und Schriftgestalter. Er betreibt die Schriftfoundry nouvellenoire.ch.

12 Der australischer Gestalter David Pidgeon leitet das Grafikbüro Design by Pidgeon.

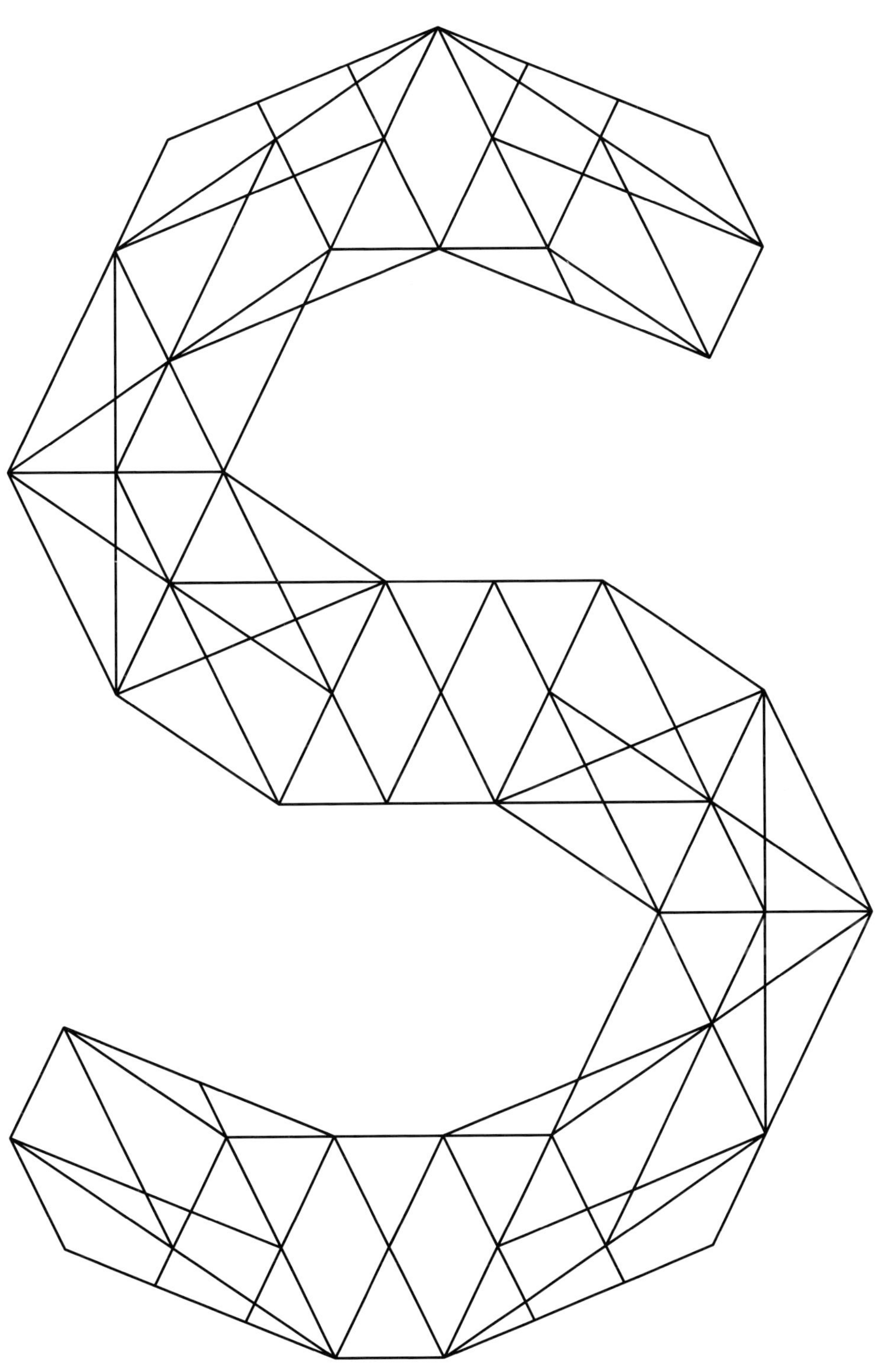

AB EIFFEL

»Gestaltung ist auch heute noch Arbeit, Erfahrung, Handwerk und Kreativleistung. Es gibt keine Abkürzungen und Schnellkurse …«

André Baldinger
→ S. 27

stilistischer/ästhetischer Natur sein, konzeptueller Natur oder beides zusammen. Ich suche die maßgeschneiderte Lösung für eine von mir oder von außen kommende Problematik zu finden. Sehe ich, dass es bereits Schriften gibt, die alle nötigen Kriterien und Qualitäten haben, und die gibt es natürlich heute für eine Vielzahl von Anwendungen, dann macht es für mich weniger Sinn, eine weitere Schrift zu entwerfen.

Welche Nutzung Ihrer Fonts von anderen Gestaltern hat Ihnen gut gefallen und wo hätten Sie diese lieber nicht gesehen?

Schriften einmal in »freier Natur«, sprich in den Händen von Benutzern, haben ihr eigenes Leben. Das muss man akzeptieren. Natürlich freut es mich immer wieder, eigene Schriften gut und gekonnt angewendet zu sehen. Die Newut wurde ebenso oft gut wie schlecht eingesetzt. Ihr konzeptueller/experimenteller Ansatz hat auch Benutzer angezogen, welche mangelnde eigene Ideen mit einem speziellen Schrifttypus wettmachen wollten.

In diesem Punkt hat sich wenig geändert. Gestaltung ist auch heute noch Arbeit, Erfahrung, Handwerk und Kreativleistung. Es gibt keine Abkürzungen und Schnellkurse, um gekonnt mit der Materie umgehen zu können.

Das Interview mit André Baldinger führten Lynn Blees und Luzia Hein.

ANDRÉ BALDINGER

André Baldinger[1] (*1963) studied with Hans Rudolf Bosshard[2] in Zurich and passed a »NDS« in type development / type design at the Atelier National de Recherche Typographique in Paris. In 1995, he founded his own studio. In 2008, he established, together with Toan Vu-Huu,[3] the Design Agency »Baldinger•Vu-Huu«. Baldinger is a member of the Alliance Graphique Internationale, developed numerous fonts, visual identities, the design of posters, books, and exhibitions as well as scenographies. He teaches type design at ZHdK,[4] EnsAD[5] Paris and other institutions and is Co-Director of EnsADLab Type. Among his fonts are AB BaldingerPro, AB BDot & AB BLine family, *AB Eiffel*, and AB Newut.

Would you please introduce yourself briefly?

ab: I was born in Zurich, grew up in Flims and in the area surrounding Lake Greifen. My father was from Zurich whereas my mother came from Leipzig. When I was 19, I moved to Zurich. My first studies covered the field of electrical engineering. I gave them up after three semesters in order to go to Robert Krügel, the former artistic director of Swiss banknotes. There I did an internship in the field of graphic arts. Later, I collaborated with Richard Feurer, Hannes Wettstein,[6] and Eclat,[7] a newly established design company. At that time typography became the favourite element of my work. I deepened my knowledge in a postgraduate degree programme for typography with Hans Rudolf Bosshard – an important and inspiring experience. Afterwards I realized that I didn't only want to use typography but that I also wanted to create it. In 1992, I submitted my application documents to the ANCT (Atelier national de création Typographique) in Paris. So I moved to Paris for my continuing education diploma in the field of type design. I met Hans-Jürg Hunziker,[8] Albert Boton[9] and Adrian Frutiger,[10] became attached to that city, and stayed there.

The Écal in Lausanne appointed me to my first professorship. Since that time I have been very engaged in graphic and type design as well as teaching. In recent years, I have increasingly dealt with Asian font systems, and I travelled to Japan, China, and Taiwan and stayed there repeatedly.

What do you find interesting about type design? Why did you decide for this field of design?

I'm interested in typography because it allows me to give a visual, conceptual, and structural expression to language and subject matters. Type is my main subject. I can use the corresponding font more precisely if I create it myself and have it individually tailored. Therefore type design is an integral part of my design practice. Moreover, I think that the design of fonts isn't only an excellent training of your visual perception and that it can show you optical principles, but it can also make you more aware of conceptual working and contexts – from the detail to the whole thing and reversely.

Do you see any typographic trends in the design of fonts and the use of these fonts?

There are always trends and movements which are linked to time and social, political, cultural, and technical conditions. The present is the result of our history.

Internet and an easy access to information make trends disappear at a quickening pace. But typefaces seem to be a bit more independent than their application. I can see a scary amount of examples which are completely interchangeable in the field of application though they are formally convincing and have certain qualities.

Innovative solutions of good quality result from intensive studies of the respective subject and its context. Superficiality is out of the question. The cultural background always influences design, too. However, the world is constantly getting smaller, more international, and for this reason, frequently also shallower, I'm afraid to say. Culture-bound aesthetics used to be more geographically localized. But now I think that there is some graphic design from England which is »more Swiss« than the design from Switzerland. I'm of the opinion that each designer should create authentic work based on his own cultural background.

Now there is an easier access to the creation of type design. Do you think that this fact influences the quality of fonts either in a negative or in a positive way?

It's certainly good that specialized literature and studies as well as increasingly better tools have made type design more accessible. – This field of design used to be only accessible to a relatively small group of people. – During the last twenty years, the production of fonts has downright

1 andrebaldinger.com baldingervuhuu.com abtypefoundry.com

2 Hans Rudolf Bosshard (*1929) is type setter, artist, designer and editor of a number of different publications, lately freelance book designer for Niggli Publishers in Switzerland.

3 Toan Vu-Huu (*1973) was born in Germany, studied in Darmstadt and works as a graphic designer in Paris.

4 The Zurich University of the Arts (ZHdK) results from the fusion in 2007 of the Academy of Design and Art (HGKZ) and the Academy for Music and Drama (HMT). The Zurich Design Museum is linked to the ZHdK (→ S. 94).

5 The EnsAD, École national des Arts Décoratifs de Paris, was founded in 1766.

6 Hannes Wettstein (1958–2008) was a Swiss industrial designer who worked, among others, for Bosch and Lamy.

7 Eclat, launched in 1988 by Daniel Zentner, is a Swiss Branding Consultant. Among their clients are SSB and Swisscom (→ S. 103).

8 Hans-Jürg Hunziker (*1938) is a Swiss type designer. He worked with Adrian Frutiger in Paris and designed the Siemens Corporate font family.

9 The graphic designer Albert Boton (*1932) worked as designer and typographer with Adrian Frutiger for Deberny & Peignot.

10 Adrian Frutiger (*1928), Swiss type designer who significantly influenced the »Swiss Typography«. Some of his most famous typefaces are the Frutiger, the Univers, the OCR-B and the Avenir.

exploded. This development has also brought forward an intensive engagement with typefaces and has spotlighted typographical questions.

Traditional type designers used to draw their designs and leave the technical work to specialists at the studios of enterprises which were engaged in the field of typefaces. Now such a handling is becoming increasingly rare, and many designers do nearly everything from A to Z. Despite this development, especially in recent years, technical devices and the corresponding knowledge have become considerably more tricky and complex with regard to Unicode and OpenType. Also the use of delta hinting for the best possible presentation on the screen requires a lot of know-how and takes a considerable amount of time if you want to do it thoroughly. However, today it's not only possible to create and realize design autonomously, but also the distribution channels have become more diverse and easier to cope with thanks to the Web.

In fact, I'm not so sure whether the quality can proportionally measure up to the quantity of fonts which are realized every year. On the other hand, there have never been as many high quality fonts at our disposal as there are now.

How did you come up with the idea of creating Eiffel? How did you find the rules for your design?

When I was looking at the Eiffel Tower and analysing its structure, the idea crossed my mind. I mainly adhered to two leading parameters: on the one hand, there was the typology of a slightly geometrical architecture, and on the other hand there were specific geometrical angles which, among other things, could be found in this structure. Since it was necessary to create a solid (area filling) text version with two font weights – in addition to the display version of this font, which was intended for the use of type at large sizes – and taking into consideration the three platform levels of the Eiffel Tower and its tapering structure, I got the idea to supplement the display version by adding two more font-weights.

So this font family has five font weights, in the solid text version Regular and Bold, in the display version Level1, Level2, and Level3. As a result, you have the following scheme from the bottom up: At ground level, there are the text versions AB Eiffel-Regular and -Bold for the bulk of texts. Then there are the platform levels with AB Eiffel-Niveau1 for the first, AB Eiffel-Niveau2 for the second, and AB Eiffel-Niveau3 for the third platform level.

What kind of problems do you face when you create a font which is constructed like Eiffel? How do you solve such problems?

There were formal, optical, aesthetic as well as technical problems. The formal and aesthetic problems had to be solved mainly for the display versions. With a defined design principle you have less leeway than in free drawing. There were hardly any models which could have served as orientation. Much time was spent on finding the ideal design.

Later, digitalization was an enormous technological challenge. With stroke widths of five units (first level,) two units (second level), and one unit (third level), I had reached my technical limits, which kept me busy for quite a while. The fine stroke widths turned out to be problematic in connection with the positioning in the fixed units of the software. The problems of the solid versions mainly referred to optical/aesthetic factors and coherence. The structure of the display version, once filled out, had to be carefully painted over by staying as close as possible to the display version and without making it look blotchy.

Anton Studer[11] collaborated with me in this tricky task during the hot summer of 2009.

Was the font that you had designed realized for the signage of the Eiffel Tower?

The font was meant for the signage system of the Eiffel Tower. However, in those days the political and strategic framework conditions were so complex that the project was repeatedly delayed, and in the end, after some political changes, it wasn't carried out anymore. But this year it will be used for a big signage project in Melbourne by my Australian colleague David Pidgeon.[12]

Is it generally important to you that a font is derived from its content or that it has an additional value which can even be more essential than aesthetic aspects?

Certainly, I always try to include the aesthetic aspect as much as possible. However, beauty and harmony aren't always the guiding principle. This can be simply boring or inappropriate under certain conditions. Context and range of use play an important role.

Coherence is another parameter. A design which hasn't a high degree of coherence is immediately worse off. In the field of fonts this parameter is particularly important.

Most of the fonts which I created so far were caused by the necessity to propose solutions for fonts which were missing or in my opinion not good enough. Usually I was interested in specific parameters. These parameters could be of a stylistic/aesthetic or of a conceptual nature or of both of them. I normally look for a tailor-made solution to a problem that I'm confronted with. If I realize that there are fonts available which meet all the necessary criteria and qualities – today this is the case for a great many applications – it certainly doesn't make much sense for me to create one more.

In which cases did you like the use of your fonts by other designers and when did you dislike it?

Once a font is »free«, which means in the hands of its users, it will lead its own life. You have to accept that. Of course I always feel happy about a good and skilful use of one of my fonts. Newut was used in an appropriate way as many times as it was inappropriate. Its conceptual/experimental approach has also attracted those users who wanted to compensate a lack of ideas with a special kind of font.

Little has changed in this regard. Even today, design requires work, experience, craftsmanship, and creative achievement. There aren't any shortcuts or crash courses which could lead to a skillful handling of this matter.

André Baldinger was interviewed by Lynn Blees and Luzia Hein.

11 Anton Studer is a Swiss graphic and type designer and founder of nouvellenoire.ch.

12 The Australian designer David Pidgeon is Director of the Studio Design by Pidgeon.

Man wird den Eindruck nicht los, dass die Medien mit ihrem Gesamtverhalten ständig an dem Ast herumsägen, auf dem sie selbst sitzen. Scheinbar gibt es daraus kein Entrinnen, weil der Wettbewerb auf dem Medienmarkt inzwischen zu einem Existenzkampf geworden ist, in dem die Wettbewerber versuchen sich gegenseitig mit größten Aufmerksamkeitserregern und höchster Aktualität zu überbieten.

Adelle Sans Regular – 7pt

Einige ziehen sich kopfschüttelnd ins Private zurück (Konsumverweigerung). Andere nehmen die »Informationen« und ihre Fragwürdigkeiten dort hin, wo sie diese zwar auch nicht verhindern, aber abschalten oder umgehen können.

Adelle SemiBold Italic – 10pt

Wir haben es überwiegend mit einem nicht-linearen dynamischen Geschehen zu tun, das sich durch interagierende innere und äußere Einflussgrößen laufend verändert.

Adelle Bold – 18pt

Wir überschätzen unsere Deutungskompetenz

Adelle Sans Bold – 32pt

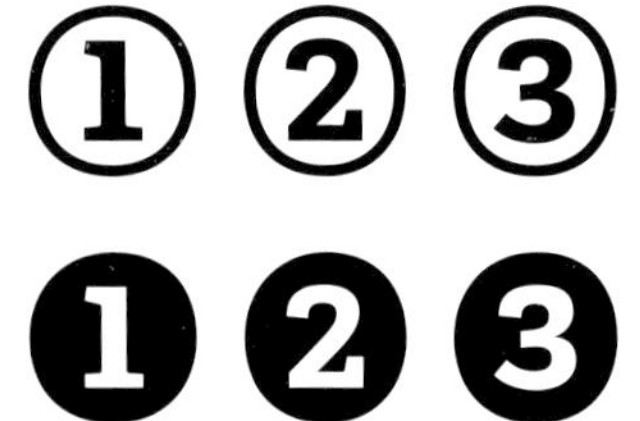

Adelle Thin, *Italic*
Adelle Light, *Italic*
Adelle Regular, *Italic*
Adelle SemiBold, *Italic*
Adelle Bold, *Italic*
Adelle ExtraBold, *Italic*
Adelle Heavy, *Italic*

Adelle Sans Thin, *Italic*
Adelle Sans Light, *Italic*
Adelle Sans Regular, *Italic*
Adelle Sans SemiBold, *Italic*
Adelle Sans Bold, *Italic*
Adelle Sans ExtraBold, *Italic*
Adelle Sans Heavy, *Italic*

Schrift. Typeface.	Adelle
Gestalter. Designer.	Veronika Burian & José Scaglione
Label. Foundry.	type-together.com
Jahr. Year.	2009

Während Adelle eine serifenbetonte Linear-Antiqua ist, die besonders für intensive redaktionelle Arbeit – hauptsächlich in Zeitungen und Zeitschriften – gedacht ist, ist sie durch ihre Persönlichkeit und Flexibilität doch eine wirklich vielseitige Schrift. Das mittelschwere Gewicht sorgt in der Textgröße für ein neutrales und lesbares Schriftbild mit der Robustheit, die man für einem Zeitungsdruck erwartet. Die unauffällige Erscheinung, Struktur und leicht dunkle Farbe sorgen dafür, dass sich die Schrift auch für durchgehende Textpassagen und selbst für anspruchvollste redaktionelle Zwecke eignet.

In ihren größeren Formen zeigt Adelle mittels besonderer Charakteristika ihre Persönlichkeit, was die Wiedererkennung leicht macht. Sie hat 14 Schnitte von leicht bis schwer, mit mehr als 800 Zeichen pro Font. Neuerdings gibt es auch für alle 14 Schnitte eine Unterstützung für kyrillische Sprachen.

While Adelle is a slab serif typeface conceived specifically for intensive editorial use, mainly in newspapers and magazines, its personality and flexibility make it a real multi-purpose typeface. The intermediate weights make for a very legible and neutral look when used in text sizes, providing the usual robustness expected in a newspaper font. The unobtrusive appearance, texture and slightly dark colour ensure a flawless performance in continuous text settings, even for most demanding editorial applications.

As it gets larger in print, Adelle shows its personality through a series of measured characteristics, making it easy to remember and identify. It has 14 styles, ranging from light to heavy, with more than 800 characters per font. Recently, support for Cyrillic script languages has been added to all 14 styles.

Zeitschrift

KULTUR

»*Citation*«

Newsflash

Feuilleton

#Actualité

Adelle Thin
Adelle Regular
Adelle Sans SemiBold Italic

Adelle ExtraBold Italic
Adelle Light
Adelle Sans Heavy

Die tschechische Designerin *Veronika Burian*[1] (*1973) studierte in Deutschland Industrial-Design und arbeitete als Produkt- und Grafik-Designerin. Im Anschluss Master-Studium Typeface-Design an der Universität Reading (GB). Burians Masterarbeit, die Schrift Maiola,[2] wurde 2004 mit dem TDC Certificate of Excellence in Type Design ausgezeichnet. 2006 gründete die Schriftgestalterin zusammen mit *José Scaglione*,[3] den sie in Reading kennengelernt hatte, die Type Foundry »TypeTogether«.[4] Zahlreiche ihrer gemeinsam gestalteten Schriften wurden im Rahmen der ED-Awards der letzten Jahre ausgezeichnet. Burian ist mit Vorträgen über Typografie auf zahlreichen internationalen Konferenzen präsent.

2012 schufen Sie eine Sans Serif-Version Ihrer bekannten Schrift Adelle. Dieser Font erfuhr kürzlich ein Update, um auf Displays besser lesbar zu sein. Was ist Ihre Meinung zu Webfonts und dazu, dass »normale« Schriften für den Gebrauch am Bildschirm optimiert werden?

vb: Wenn Sie mit Schriften Desktop-Fonts meinen, dann denke ich, dass die meisten von ihnen soweit verändert und optimiert werden können, dass sie sich für den Bildschirm eignen. Die Frage ist natürlich, wofür die Schrift verwendet wird, soll es für Überschriften, für Untertitel oder für Fließtext sein? Texte mit mehr als 24px brauchen nicht so viel Aufmerksamkeit wie kleine Textgrößen, wenn es um Bildschirm-Optimierung geht. Es liegt in der Natur der Pixel – da Sie viereckig sind – und an ihrer Auflösung auf dem Bildschirm, dass gewisse Designeigenschaften wie flache Kurven oder kleine Rundungen Probleme bereiten.

Webfonts bieten dem Designer viele Möglichkeiten, aber genau wie bei Desktop-Fonts muss man sich genau überlegen, welche man einsetzt.

Wie ist es für Sie, wenn Sie Ihre Schriften »in Aktion« sehen? Haben Sie jemals festgestellt, dass sie für etwas verwendet wurden, womit Sie nicht einverstanden waren?

Ich freue mich natürlich, wenn meine Schriften intelligent eingesetzt werden. Es macht Spaß, ihnen draußen in der Alltagswelt zu begegnen. Schriften sind da, um benutzt zu werden, idealerweise zu dem Zweck, für den sie geschaffen wurden, aber ich lasse mich auch gern überraschen. Zum Beispiel hatte ich Bree[5] nie als eine Schrift für Zeitungen gesehen, aber jetzt

1 type-together.com

2 Maiola ist eine Antiqua, die von tschechischer Schriftgestaltung und handschriftlicher Kalligrafie beeinflusst wurde. Sie wurde 2005 von TypeTogether veröffentlicht.

3 José Scaglione (*1974) ist ein argentinischer Grafik- und Type-Designer, der an den Universitäten in Rosario und Buenos Aires in Argentinen unterrichtet. Darüber hinaus ist er seit 2007 Mitglied des Vorstandes des »Association Typographique Internationale«. Er gründete 2006 zusammen mit Veronika Burian die Type Foundry TypeTogether (→ S. 60, 126).

4 Nachdem sich José Scaglione und Veronika Burian während ihres Master-Studiums an der University of Reading kennengelernt hatten, gründeten sie 2006 die Foundry »TypeTogether« (→ S. 59).

5 Bree (2009) ist eine Groteske, die von Handschrift beeinflusst ist. Sie wurde für den Einsatz als Headline-Typeface oder Branding-Schrift optimiert.

habe ich sie schon einige Male sehr erfolgreich auf diese Weise angewandt gesehen.

Wie wissen Sie, wann eine Schrift »fertig« ist? Ist es eine bewusste Entscheidung, wann Sie aufhören, sie zu verbessern, oder lassen Sie sich von Ihrem Instinkt leiten?

Man könnte sagen, dass eine Schrift nie ganz »fertig« ist, und es passiert schon, dass ich immer noch etwas verbessern möchte, auch wenn der Font schon fertig ist. Doch meist gibt es einen Moment im Designprozess, wenn es sich »richtig« anfühlt, also vermute ich schon, dass ich mich von meinem Instinkt leiten lasse.

Was sind Ihre persönlichen Kriterien für eine »gute« Schrift?

Die Qualität der Zeichnung und Einheitlichkeit im Design, die sich in den Details zeigt, die sich immer wiederholen. Ferner Zweckmäßigkeit, was den Leser betrifft, und Originalität im Ausprobieren von Ideen.

Viele Ihrer Schriften beschränken sich nicht auf das lateinische Alphabet – ist es Ihnen wichtig, dass Ihre Schriften »global players« werden, indem Sie das Spektrum auf andere Alphabete ausweiten?

Nun ja, es ist schon wichtig, dass man nicht kurzsichtig wird und sich zu sehr eingrenzt. Und es ist eine interessante Herausforderung, sich mit anderen Schriftsystemen zu befassen.

Wie fangen Sie an, wenn Sie eine neue Schrift entwickeln? Viele Ihrer Schriften beziehen sich auf die Kalligrafie: arbeiten Sie lieber analog oder digital?

Es ist wichtig, dass man zuerst die grundsätzlichen Proportionen definiert. Es gibt die sogenannten Kontrollbuchstaben, n, o, H, O, dann muss man die Ober- und Unterlängen festlegen, dafür benutze ich das p und das h. Außerdem entwerfe ich das c und das y. Wenn es eine Auftragsarbeit ist, muss ich manchmal schon im Anfangsstadium ganz spezielle Buchstaben berücksichtigen, je nach den Wünschen des Kundens. In Reading, wo ich den Masterkurs machte, nahmen wir »adhesion« als Testwort, an dem wir unsere Ideen ausprobierten, aber ich selbst habe kein solches Wort. Stattdessen teste ich den Font in einem unsinnigen Text, in dem die Buchstaben in den verschiedensten Kombinationen aufeinander treffen.

Ich glaube, Zeichnen ist anfangs noch immer die beste Art und Weise, um Ideen zu entwickeln. Obwohl ich darin nicht besonders gut bin, hilft es mir, über Designmerkmale zu entscheiden wie Terminals, Serifen, Kontraste usw.

Danach gehe ich direkt zum Computer. Ich persönlich halte es für Zeitverschwendung, meine Zeichnungen zu digitalisieren, aber jeder Designer hat seine eigene Methode. Bei einem Revival-Design macht es schon Sinn, oder wenn ein Buchstabe oder irgendein Detail in der Zeichnung besonders gut gelungen ist.

Als nächsten Schritt zeichne ich einen größeren Zeichensatz, vor allem die Minuskeln, damit ich mir den Font in einem Text ansehen kann. Das ist die beste Art zu prüfen, ob die Buchstaben miteinander harmonieren und wie sie auf der Druckseite wirken.

Die Punktgröße zum Beispiel hängt davon ab, welchen Zweck die Schrift erfüllen soll. Will man es für ein Display, soll damit ein Buch gesetzt werden, soll es ein Grobdruck werden oder ist es für eine Zeitung? Idealerweise prüft man das Schriftbild in dem Umfeld, für das es bestimmt ist, und in verschiedenen Punktgrößen.

In dieser Phase achte ich auch bereits auf die Laufweite, denn die beeinflusst das Erscheinungsbild der Buchstaben ganz entscheidend. Bei größeren Drucken sehe ich mir auch die Umrisse der Buchstaben an. Dazu ist ein guter Drucker äußerst wichtig. Wenn ich mit den Minuskeln zufrieden bin, fahre ich mit dem restlichen Zeichensatz fort, wobei ich den beschriebenen Prozess viele Male wiederhole.

Problematische Zeichen sind s, y, z und die 2, aber dazu habe ich auch keinen wirklichen Trick, außer zu empfehlen, dass man sich historische Schriften, Kalligrafien und verwandte optische Beispiele ansieht, denn die Formen unserer Buchstaben sind durch das Schreiben mit der breiten Feder entstanden. Die Grundstrukturen und ihre Abwandlungen sind ziemlich festgelegt. Selbst bei einer geometrisch-konstruierten Schrift ist es von Vorteil, wenn der Designer mit diesen Grundsätzen vertraut ist.

Der klassische Ansatz ist, dass man mit Normal anfängt, denn das war der am häufigsten verwendete Schnitt, häufiger als fett, kursiv usw. Heute allerdings, besonders bei den

Sans Serif-Schriften, kann es vernünftiger sein mit Light oder Extra Bold anzufangen und dann die anderen Gewichte zu interpolieren. Ich hatte auch schon Projekte, bei denen ich mit einer Heavy anfing und die anderen Gewichte von ihr aus entwickelte ohne zu interpolieren. Auch das hängt wieder sehr von den Anforderungen der Kunden ab.

Was inspiriert Sie eine neue Schrift zu schaffen?

Die Inspiration kann von überall herkommen: von einem Schild, einem Buch, einer Beschriftung, Graffiti, einer Handschrift, alten Exemplaren oder etwas Volkstümlichem. Ideen hängen auch sehr von den besonderen Umständen eines Projekts ab, sie können von kommerziellen Interessen und Bedürfnissen diktiert werden, von persönlichen Interpretationen historischer Vorbilder oder von einer interessanten Beschriftung. Auf jeden Fall wird die Idee natürlich auch von der beabsichtigten Verwendung beeinflusst und von dem Ergebnis, das man sich davon erhofft, z.B. wird ein Design für einen Zeitungsfont sehr anders aussehen als ein Font für den Gebrauch am Bildschirm.

Warum gründeten Sie Ihren eigenen Schriftvertrieb? Welche Vor- und Nachteile ergeben sich aus der Verantwortung eines eigenen Betriebs?

Ich habe mich als Angestellte nie sehr wohl gefühlt und wollte mein eigenes Ding machen, meinen eigenen Ideen nachgehen. Eine eigene Foundry zu haben bedeutet, dass man sämtliche Entscheidungen selbst treffen kann und unabhängig ist. Allerdings bedeutet es auch, dass man sehr fleißig sein und eine gewisse Unsicherheit in Kauf nehmen muss.

Ihre Foundry »TypeTogether« ist ein virtuelles Studio: Sie und Ihr Partner José Scaglione leben in zwei verschiedenen Ländern – welche Auswirkungen hat das auf Ihren Arbeitsfluss? Wie stellen Sie sicher, dass die Qualität Ihrer Arbeit konsistent bleibt, selbst wenn Sie nicht persönlich zusammenkommen?

José und ich haben ein sehr gutes Verhältnis, das auf Vertrauen und gegenseitigen Respekt aufgebaut ist. Wir diskutieren viel über Skype, so ziemlich jeden Tag. Es ist nicht wirklich erforderlich, dass man jeden Tag zusammen im selben Studio verbringt. Wir treffen uns zweibis dreimal im Jahr, dann arbeiten wir allerdings sehr intensiv zusammen.

Wie schwer ist es zu überleben, wenn man so leicht Raubkopien von Fonts machen kann?

Es ist schwer zu sagen, wie viel Einkommen wir durch Raubkopien verlieren. Aber ich bin überzeugt, dass man die Menschen erziehen muss, damit sie wissen, wie viel Arbeit es bedeutet, eine Schrift zu entwickeln. Man muss mit den Studenten reden und auch in der Öffentlichkeit ein Bewusstsein für die Qualität und den Wert guter Typografie schaffen. Wir sind keine Schriftenpolizei, aber wir erwarten gegenseitigen Respekt.

Ist Ihnen bewusst, dass Ihr Arbeitsfeld von Männern dominiert wird?

Eigentlich nicht, und es wechselt ja auch stark. Viele junge Frauen werden professionelle Schriftdesignerinnen oder haben zumindest ein großes Interesse daran. An meinen Workshops nehmen immer viele Frauen teil.

Da Sie in einem Team arbeiten, wie wichtig ist es Ihnen, über Ihre Ideen, Entwürfe und Zeichnungen die Kontrolle zu behalten?

Ich glaube, ich möchte schon eine gewisse Kontrolle behalten, aber mit José habe ich ein gutes Verhältnis des Vertrauens und des Respekts. Wir teilen uns unsere Arbeit ziemlich gleich und haben kein Problem damit, uns die Urheberschaft einer Schrift zu teilen, auch wenn einer einmal »mehr Arbeit« als der andere daran geleistet hat. Ich arbeite gern im Team.

Wie fängt Ihr Tag an? Welches Umfeld brauchen Sie, um kreativ zu sein? Wie sieht Ihr Schreibtisch aus?

José und ich haben einen Grundsatz: Ohne Kaffee keine Fonts! Ich glaube, ich brauche eine Art kreatives Chaos, verschiedene Bilder, Bücher, Geräte, Erinnerungsstücke usw. Mein Schreibtisch ist schon ziemlich unordentlich.

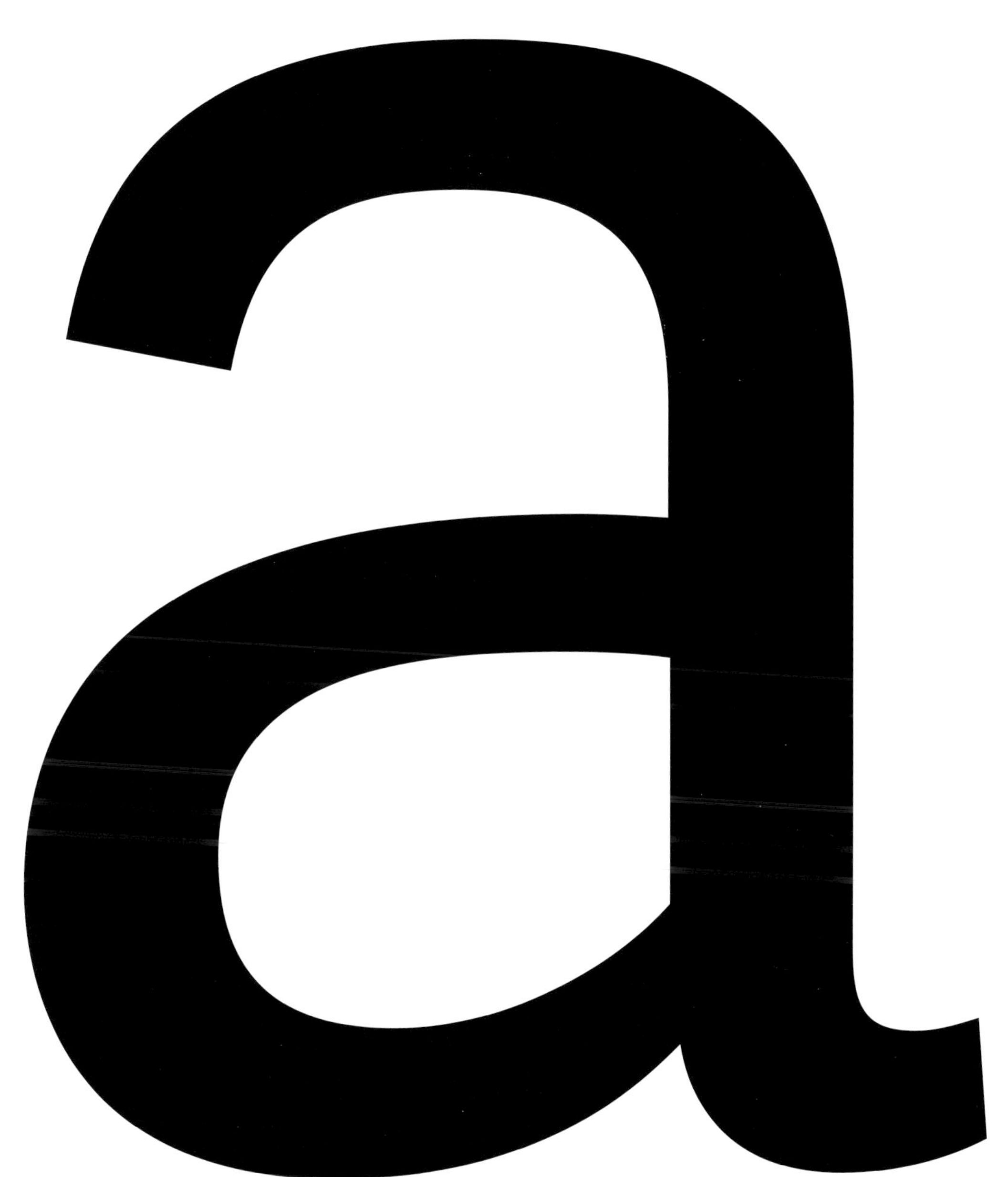

ADELLE

Wie halten Sie sich mit Trends auf dem Laufenden? Ist Ihnen der Zeitgeist wichtig?

Nun ja, man muss schon darauf achten, was gerade passiert, aber unser Geschäft sind nicht diese angesagten Reklameschriften, die ein paar Monate lang »in« sind und dann in Vergessenheit geraten. Wir sind viel mehr daran interessiert, Problemlösungen zu finden und langlebige Designs zu schaffen. Natürlich bemühen wir uns, zeitgemäß zu sein. Von Revivals halte ich allerdings nicht sehr viel.

Gibt es im Moment irgendwelche typografische Trends, die Ihnen auffallen?

Ich weiß nicht, ob die jetzt im Trend sind, aber ich denke, sie könnten ihre Liebhaber finden: rückwärtsliegende Kursivschrift und hochkontrastierende Sans Serifs.

Gibt es irgendwelche Schriften oder Schriftdesigner, die Sie so sehr beeindruckt haben, dass es Ihre Sicht auf Schriften verändert hat?

Natürlich gibt es zeitgenössische Schriften von Kollegen, die ich bewundere und respektiere. Die beiden Schriftdesigner, die mich am Anfang am meisten beeinflusst haben, waren die tschechischen Designer Vojtěch Preissig[6] und Oldřich Menhart.[7] Beide arbeiteten in der ersten Hälfte des 20. Jahrhunderts.

Haben Sie irgendwelche Ratschläge für frischgebackene Schriftdesigner? Gibt es etwas, was sie beachten sollten?

Sie sollten sich viele gute Beispiele ansehen; sozusagen zu den Wurzeln zurückkehren, statt nur die digitalisierten Versionen von Klassikern wie Garamond[8] oder Bodoni[9] anzustarren. Die Originale können euch viel mehr zeigen. Besucht Museen, die darauf spezialisiert sind, wie zum Beispiel das Museum Plantin-Moretus[10] in Antwerpen; nehmt an Kalligrafiekursen teil; lest Bücher über die Geschichte der Schrift und versucht, euch so viele verschiedene Schriftbeispiele anzusehen wie möglich.

Das Interview mit Veronika Burian führten Alexa Spors und Lisa Steinhauer.

6 Vojtěch Preissig (1873–1944) war ein tschechischer Typograf, Designer, Illustrator und Maler. Er studierte »Industrielle Kunst« und »Dekorative Architektur« in Prag. Seine Preissig Antiqua beeinflusste die Gestaltung von Schrift und Printmedien im frühen 20. Jahrhundert.

7 Oldřich Menhart (1897–1962) war ein tschechischer Buch- und Schriftgestalter, der für die Staatsdruckerei Prag arbeitete und freiberuflich Druckschriften für Schriftgießereien wie die Bauersche Gießerei (Frankfurt am Main) oder die Monotype Corporation (London) entwickelte.

8 Garamond ist die Bezeichnung für eine Gruppe von Schriften, die durch Claude Garamond (1480–1561) geprägt wurde. Sie werden als humanistische Antiquas klassifiziert, die über eine sehr gute Lesbarkeit in Printmedien verfügen.

9 Giambattista Bodoni (1740–1813) hat eine Reihe klassizistischer Antiquaschriftarten geschaffen, die als Bodoni bezeichnet werden. Sie verfügen über einen hohen Strichstärken-Kontrast und flache Serifen (→ S. 135).

10 Plantin-Moretus (Antwerpen/B) war eine Druckerei des 16. Jahrhunderts, die zu einem Museum für Buchdruckkunst und Typografie wurde.

The Czech designer *Veronika Burian*[1] (*1973) studied industrial design in Germany and worked as product and graphic designer. Then she successfully participated in the Typeface Design Master's Programme of the University of Reading (Great Britain). Burian's Master's thesis, the Maiola[2] font, was awarded the TDC Certificate of Excellence in Type Design in 2004. In 2006, Veronika Burian established with *José Scaglione*,[3] whom she had met in Reading, the »TypeTogether«[4] type foundry. Their fonts received several awards at the ED-Awards competition over the past years. Burian gives lectures on typography in numerous international meetings.

In 2012 you released a sans-serif version of your popular typeface Adelle. This font was recently updated in order to achieve a better legibility on displays. What are your thoughts on webfonts and about optimizing »regular« typefaces for the use on screens?

vb: If by typefaces you mean desktop fonts, I think that most typefaces can be tweaked and optimized to be used on screen. The question is in which function the typeface will be used, is it for big titles, subtitles or continuous text? Texts above 24px won't need as much attention with regards to on-screen optimization as small text sizes. Due to the nature of pixels, i.e. squares, and the resolution of screens, certain design features like shallow curves or small round corners will create problems.

Webfonts offer a great opportunity for designers, but as with desktop fonts, thought has to be given to the selection of appropriate fonts.

How do you feel when you see your typefaces »in action«? Have they ever been used in a context or project that made you feel uncomfortable or that you disapproved of?

I like to see my typefaces used well, of course. It is fun to see them out in the world. Typefaces are there to be used, ideally in the way for which they were intended, but I am also quite happy to be surprised. For example: I never thought of Bree[5] as a newspaper face, but then I saw it used really well several times.

How do you know that a typeface is »done«? Is it a conscious decision to stop tweaking it or do you go with your instincts?

You could say that a typeface is never done and it does happen that I feel like changing details after the font was completed. There comes a point in the design process when it feels right, so I guess you can say I go with my instinct.

What are your personal criteria for a »good« typeface?

Quality of drawing, consistency of design, which manifests itself in repeating details and features, functionality with respect to the reader, originality in exploring ideas.

A lot of your typefaces are not limited to the Latin alphabet only – is it important for you to turn your typefaces into »global players« by extending their spectrum in terms of alphabets?

Well, I think it is important not to get short-sighted and limit oneself. It is an interesting challenge to dive into other writing systems.

When designing a typeface, how do you start? A lot of your typefaces have references to calligraphy: Do you prefer to work analog or digitally?

It is important to define your basic proportions first. There are the so-called control characters, n, o, H, O, and you need to define ascenders and descenders, so I use p, h. In addition, I also design a, c, y. Very often though, when it comes to commissioned work, I need to include particular letters into the initial concept stage, depending on the client's requirement. In Reading at the MA course, we used »adhesion« as the initial test word, but I find myself not using any at all. Instead I test the font in a non-sense text where the letters come together in different combinations.

Sketching is still the best way of developing initial ideas, I think. Although I am not particularly good at it, it helps me to decide about design features, such as terminals, serifs, contrast, etc.

From there I go directly to the computer. Personally I find it a waste of time to digitize my drawings, but every designer has his/her own way. It surely makes sense for a revival design, or if a letter or detail turned out particularly well in the sketch.

In the next step, I draw a bigger character set, mainly the lowercase set to be able to test the font in a text setting. This is the ultimate way to see if the letters work well together and how their feel or impact is on the page.

The point size, for instance, depends on the purpose for the typeface. Is it for display, book setting, coarse printing or newspaper, etc.? Ideally, the typeface is tested in the environment it is designed for and in a range of point sizes.

Already at this stage, I tend to pay attention to spacing, because it greatly influences the appearance of the letters. I also look at the quality of the outlines in bigger print-outs. A good crisp printer is of utmost importance here. Once I am happy with the lowercase, I continue with the rest of the character set and repeat the process many times.

1 type-together.com

2 Maiola is a Serif influenced by Czech type design and calligraphy. It was published in 2005 by TypeTogether.

3 José Scaglione (*1974) is an Argentinian graphic and type designer teaching at the Universities of Rosario and Buenos Aires. Since 2007 he has also been a Board Member of the Association Typographique Internationale. In 2006 he founded, together with Veronika Burian, the type foundry »Type Together« (→ S. 63, 129).

4 José Scaglione and Veronika Burian met during their Master Programme at Reading University and together established the foundry »TypeTogether« in 2006 (→ S. 63).

5 Bree (2009) is a sans-serif influenced by handwriting. It was optimized for use as a typeface for headlines or as a branding script.

»There comes a point in the design process when it feels right ...«

Veronika Burian
→ S. 37

Problem characters, I would say, are the s, y, z, 2, but I don't really have any genuine tricks, other than that I recommend studying historic typefaces, calligraphy and related optical phenomena, because our letter shapes are derived from writing with a broad-nib pen. The basic structure and modulation are pretty much fixed and defined. Even a geometric and constructed typeface will benefit from the designer's understanding of those principles.

The classic approach is to start with the Regular, because it used to be the most frequently used style, more than the Bold, Italic … Nowadays, though, especially with sans-serifs, it can be more useful to start with a Light and Extrabold and then interpolate the other weights. I also had projects where I started with a Heavy and developed the other weights from there without interpolation. Again, it often depends on the client's requirements.

What inspires you to create a new typeface?

Inspiration can be anything: a sign, a book, lettering, graffiti, handwriting, old specimens, vernaculars, etc. Ideas depend on the circumstances of the project, but they can be driven by commercial need and interest, personal interpretation of historic models or some interesting lettering. In any case, the idea is also influenced by the intended use and the requirements that follow from this, e.g. the design concept for a newspaper font will look very different from a font for onscreen use.

Why did you start your own type foundry? What advantages and disadvantages come with the responsibility of being your own boss?

I never liked being employed and I wanted to do my own stuff, explore my own ideas. Having your own foundry means you make all decisions by yourself and you are independent, however it also means hard work and uncertainty.

Your foundry »TypeTogether« is a virtual studio: You and your partner José Scaglione live in two different countries – how does this affect your workflow? What kind of routines do you use to ensure that the quality stays the same, even if you cannot see each other in person?

With José we developed a very good relationship of trust and respect. We discuss a lot over Skype, pretty much every day. It is not really necessary to be in the same studio all the time. We work intensely when we do get together two or three times a year.

How hard is it to make a living at a time when typeface can easily be pirated?

It is hard to say how much revenue we actually lose by font piracy. I believe in educating people about the amount of work it takes to design a text typeface, speaking to students, making the general public aware of the quality and value of good typography. We are not the font police, but we do ask for mutual respect.

Do you notice that you are working in a male-dominated field of work?

Not really, and actually, it is changing a lot. Many young women are becoming professional type designers or at least have a strong interest in it. I always have a lot of women in my workshops.

Since you are working in a team, how important is it for you to be in control of your concepts, drafts or ideas?

I probably like to have some control, but with José, we developed a very good relationship of trust and respect. We share the amount of work pretty much equally and have no issues with crediting our typefaces to both of us, even if one of us did »more work« than the other. I like working in a team.

How do you start your day? What kind of environment do you need in order to be creative? What does your desk look like?

We have a saying with José: No coffee, no fonts! I guess I need some kind of a creative chaos, exposure to different pictures, books, gadgets, memorabilia etc. My desk is actually pretty untidy.

How do you keep up with trends? Is Zeitgeist something you care about?

Well yes, we do need to keep an eye on what's happening, but we are not in the business of trendy display faces that are hip for a few months and then die. We are much more interested in solving problems and creating long-lasting designs. We do aim to be contemporary though. I don't like doing revivals very much.

Are there any particular typographic trends that catch your eye at the moment?

Not sure if these are already trendy, but I think they might catch on: reverse italics and high contrast sans-serifs.

Are there any typefaces or type designers that impressed you so much that they changed the way you see type now?

There are certainly contemporary typefaces by colleagues that I admire and respect. The two designers that influenced me the most, when I started out, were the Czech designers Vojtěch Preissig[6] and Oldřich Menhart.[7] Both were active in the first half of the 20th century.

Do you have any advice for »newbie« type designers? Is there anything they should look out for?

They should look at good old examples; go back to the roots so to speak and not just stare at the digitized versions of classics like Garamond[8] or Bodoni.[9] The originals will show you much more. Visit specialized museums like the Museum Plantin-Moretus[10] in Antwerpen; take some calligraphy courses; read books about type history and try to look at as many type specimens as possible.

Veronika Burian was interviewed by Alexa Spors and Lisa Steinhauer.

6 Vojtěch Preissig (1873–1944) was a Czech typographer, designer, illustrator and painter. He studied »Industrial Art« and »Decorative Architecture« in Prague. His Preissig Antiqua influenced the design of type and print media early in the 20th century.

7 Oldřich Menhart (1897–1962) was a Czech book and type designer who worked for the Government Printing Works in Prague as well as doing freelance work, developing fonts for type foundries like the Bauersche Gießerei (Frankfurt am Main) or the Monotype Corporation (London).

8 Garamond is the name for a group of typefaces influenced by Claude Garamond (1480–1561). They are classified as humanistic Antiquas, affording very good legibility in print media.

9 Giambattista Bodoni (1740–1813) has created a number of classicistic Antiqua types, known as Bodoni. They have a high contrast in stroke widths and flat serifs (→ S. 139).

10 The Museum Plantin-Moretus in Antwerp (B) was a print shop in the 16th century which is being used as a Museum for Printing and Typography.

SIBYLLE HAGMANN

THE MIND IS LIKE AN UMBRELLA — IT FUNCTIONS BEST WHEN OPEN.

Axia Stencil Black – 40pt

A modern, harmonic and lively architecture is the visible sign of an authentic democracy.

Axia Bold – 28pt

The ultimate goal of all visual artistic activity is construction! Architects, painters and sculptors must learn again to know and understand the multi-faceted form of building in its entirety as well as its parts. Only then will they of their own accord fill their works with the architectonic spirit they have lost in the art of the salon. Let us establish a new guild of craftsmen without the pre-sumption of class distinctions building a wall of arrogance between craftsmen and artists. Together let us call for, devise and create the construction of the future, comprising everything in one form: architecture, sculpture and painting.

Axia Regular – 6pt

Art itself cannot be taught, but craftsmanship can. Architects, painters, sculptors are all craftsmen in the original sense of the word. Thus it is a funda-mental requirement of all artistic creativity that every student undergo a thorough training in the workshops of all branches of the crafts.

Axia Light – 10pt

Schrift. Typeface. Axia
Gestalter. Designer. Sibylle Hagmann
Label. Foundry. kontour.com
Jahr. Year. 2003–2012

Axia entstand 2012 aus Schriften, die ursprünglich für die Rice University School of Architecture gestaltet wurden. Sie ist eine robuste Sans Serife-Schrift mit kompakten Formen. Sie kommt in zehn Gewichten von Hell bis Schwarz mit erweiterter Sprachunterstützung, einer großen Auswahl an OpenType-Features einschließlich Kapitälchen, mehrfachen Schriftstilen und vieles mehr. Weitere Optionen der typographischen Palette sind Stencil-Display-Versionen mit einem besonderen Ausdruck, perfekt für prominente Schriftgrößen. Die Stencil-Formen sind gekennzeichnet durch freischwebende Teile, die das Auge fesseln und im Verbund miteinander eine wohlproportionierte Schrift ergeben. Von Grund auf entwickelt, öffnen die inneren Bögen, die vom Stamm ausgehen – z.B. bei den Minuskeln n oder d – stufenweise die Form und geben dem gesamten System eine konzeptionelle Klarheit. Diese Eigenschaft sorgt bei den schwereren Gewichten für eine gute Lesbarkeit, gleichzeitig trägt sie zur Vielseitigkeit der verschiedenen Gewichte bei.

Axia, designed in 2012, originates from a typeface design for the Rice University School of Architecture. The family is a robust sans serif of concise letter forms. It comes in ten weights from Light to Black with extended language support, a host of OpenType features including Small Caps, multiple figure styles, and more. Further distinctive options to the typographic palette are offered by stencil display weights with unique aesthetics, perfect for captivating type sizes. The stencil display weights consist of abstract floating parts that catch the eye and form nicely proportioned type when united.

Orchestrated from scratch, the inner arched strokes off the stem on the lowercases »n« or »d«, for example, progressively open the letter forms and express conceptual clarity throughout the system. This feature makes for better legibility in the heavier weights, while at the same time contributing to the versatility of individual weights.

100% uniwidth

100% uniwidth

100% uniwidth

100% uniwidth

100% uniwidth

100% uniwidth

100% uniwidth

100% uniwidth

Axia Stencil Black
Axia Black Italic

Axia Bold
Axia Bold Italic

Axia Regular
Axia Regular Italic

Axia Stencil Light
Axia Light Italic

Die in der Schweiz aufgewachsene *Sibylle Hagmann*[1] (*1965) schloss 1989 ihr Studium an der Basel School of Design ab und ging für den Master 1996 nach Amerika, wo sie am California Institute of the Arts ihre Leidenschaft für Typografie und Type Design entdeckte. Mit ihren Schriften hat sie zahlreiche Preise gewonnen, etwa den Swiss Federal Design Award für die Schriftfamilie Odile.[2] Hagmann hat an mehreren Schulen in Kalifornien unterrichtet und lehrt zurzeit an der University of Houston, schreibt für »Sage Publications«, die »Typografischen Monatsblätter« und »IDEA Magazine«. Sie lebt in Houston, wo sie 2000 ihr eigenes Studio »Kontour«[3] gegründet hat, über das sie seit 2012 auch ihre eigenen Schriften vertreibt.

Warum entschieden Sie sich, Ihren Master in den Vereinigten Staaten zu machen? Und warum sind Sie dort geblieben?

sh: Es schien mir aus mehreren Gründen eine gute Idee. Erstens suchte ich nach einer völlig anderen Perspektive, was Grafik-Design und Typografie betrifft, und zweitens wollte ich Arbeitserfahrung im Ausland sammeln. Angebote zum Unterrichten führten dann dazu, dass ich blieb, und nach ein paar Jahren hatte ich mich auch an die viele Sonne gewöhnt.

Gehörte die Beschäftigung mit Schriftdesign zu Ihrem Master-Programm?

Nicht direkt, zumindest war es nicht so beschrieben. Zu der Zeit gab es wenige MFA Programme, wenn überhaupt, die einen Masterkurs in Schriftdesign anboten. Jeffery Keedy[4] (oder Mr. Keedy, wie er sich nannte), der Entwickler der Keedy Sans,[5] einer viel diskutierten Schrift Anfang der 1990er Jahre, war Mitglied der Fakultät im Programm Graphic Design am CalArts (California Institute of the Arts, Valencia, Kalifornien). Es war mein größter Wunsch, im Graphic-Design-Masterkurs von ihm zu lernen, wie man digitale Schriften entwirft und entwickelt. Und zum Glück erfüllte sich dieser Wunsch, als er während meines zweiten Jahres an der CalArts einen Kurs darüber anbot.

Wie gehen Sie vor, wenn Sie eine neue Schrift entwerfen? Nehmen wir zum Beispiel Axia. Fing das mit der Form eines bestimmten Buchstabens an, von der die anderen Buchstaben dann abgeleitet wurden?

Die Wurzeln der Axia gehen zurück zum Alphabet von TwinCities, das ich 2003 für einen Wettbewerb des Design Institute der Universität von Minnesota entworfen hatte. Das Institut lud Schriftdesigner ein, sich mit der Frage zu beschäftigen, wie man mit einer Schrift das Einmalige einer Stadt ausdrücken kann. Bei dieser Arbeit experimentierte ich mit Buchstaben-

1 kontour.com

2 Odile ist eine lebendige Textschrift von Sibylle Hagmann. Als Inspiration dienten ihr Schriftexperimente von William A. Dwiggins aus dem Jahr 1937. Odile wurde 2006 mit dem Swiss Federal Design Award ausgezeichnet.

3 Kontour ist Sibylle Hagmanns 2000 gegründetes Studio mit Fokus auf Type Design und Sitz in Houston, Texas.

4 Jeffery Keedy (*1957) ist ein amerikanischer Grafik- und Type Designer, dessen Essays u. a. in Eye, I.D. und Emigre erschienen sind. Er ist bekannt für seine Schrift Keedy Sans.

5 Keedy Sans ist eine 1989 von Jeffrey Keedy gestaltete experimentelle Serifenlose. Sie wird durch Emigre Fonts vertrieben.

formen, die ein zeitgemäßes, räumliches und perspektivisches Schriftbild ergeben sollten, das lokale architektonische Besonderheiten widerspiegeln sollte. Diese Serie von Schriften blieb dann ziemlich lange liegen, aber ich wusste, dass ich damit noch nicht fertig war.

Ich nahm das Projekt wieder auf, als ich 2011 von der Rice School of Architecture den Auftrag für ein neues Logo und ihre neue Identität bekam. Das Logotype entwarf ich aus überarbeiteten TwinCities Buchstaben. Als sich die Schriften im Gestaltungsprozess weiterentwickelten, zeigte sich die Schule daran interessiert, die Fonts auch für ihre Drucksachen und ihr Webdesign zu verwenden. Nach einigen Überarbeitungsgängen wurde die RSA-Schriftfamilie der Vorgänger von Axia. Mit der Axia hatte ich außerdem noch die Idee, eine Schriftfamilie zu entwickeln, die sich dadurch auszeichnet, dass Textlängen homogen sind, unabhängig vom Gewicht. Texte, die in Leicht oder Schwarz gesetzt sind, nehmen unabhängig vom Gewicht gleichviel Raum ein. Innerhalb dieser Gegebenheiten experimentierte ich mit den inneren Bögen, die z.B. bei den Minuskeln n und d vom Grundstrich ausgehen, um die Buchstabenformen bei zunehmendem Gewicht offen zu halten. Das könnte man zum Beispiel eine charakteristische Formgebung nennen, die bei einer Serie von anderen Minuskeln angewandt wurde, und sie wurde ein charakteristisches Merkmal dieser Schrift.

Sie entwickelten RSA, den Vorgänger von Axia, also für die Rice School of Architecture. Steht hinter Ihren Schriften immer ein Auftraggeber?

Manchmal, aber längst nicht immer, ich entwickle auch gern Schriften aus eigenem Interesse. Es ist ein Prozess, der mir Spaß macht, denn das fertige Produkt ist selten das Resultat aus verschiedenen Meinungen. Bei Grafik-Design-Aufträgen kochen oft zu viele Köche mit. Man bekommt dann zwar auf demokratische Weise ein Ergebnis, aber es muss nicht unbedingt das Beste sein. Mit Schriftdesign ist das anders, hier ist der Kunde oder der Mitarbeiter weit weniger direkt involviert.

Wie stark ist der Unterschied zwischen der ersten Zeichnung einer Schrift und der endgültigen Version?

Das hängt ganz von der Schrift und den konzeptionellen Ideen ab, die dahinter stehen. Im Schriftdesign handelt es sich ja im Allgemeinen um winzige Änderungen, und doch können sie visuell einen Riesenunterschied ausmachen. Vielleicht kann man sagen, je komplexer eine Schrift in ihrer Formgestaltung ist, desto größer ist der Unterschied zwischen den Anfangszeichnungen und der endgültigen Version.

Wie wissen Sie, wann eine Schrift fertig ist?

Eine Schriftfamilie ist fertig, wenn ich mit jedem Detail zufrieden bin, einschließlich aller formellen und technischen Aspekte. Manchmal denke ich, dass ich mit einer Schrift fertig bin, nur um es mir dann doch wieder anders zu überlegen, und manchmal kommt dann eine weitere ganze Familie hinterher. Aber im Allgemeinen sehe ich es als fertig an, wenn ich wegen der Details nicht mehr den geringsten Zweifel habe. Wenn ich morgens aufwache und keine lange Liste mehr im Kopf habe, die sich mit dieser Schrift befasst – das kann ein Zeichen sein, dass sie fast fertig ist. Normalerweise überstürze ich die Herausgabe einer Schrift nicht, denn das ist etwas, das über eine gewisse Zeit hinweg reifen muss. Im Laufe der Jahre verlasse ich mich immer mehr auf meine innere Wahrnehmung und Objektivität bei der Entscheidung, ob eine Schrift fertig ist.

Cholla[6] kam 1999 heraus und wird von Emigre[7] vertrieben. Warum beschlossen Sie, Ihr eigenes Label »Kontur« zu gründen?

Als ich mich entschloss, mich mehr auf Schriftdesign zu konzentrieren, war es ein logischer Schritt, Ende 2012 meinen eigenen

6 Die Schriftfamilie Cholla (1998–1999) wurde ursprünglich für das Art Center College of Design in Pasadena gestaltet. Sie gewann 2000 den Award für Excellence in Type Design des Type Directors Club, New York.

7 Emigre Inc. ist ein Designstudio und war eine der ersten Independent-Foundries. Emigre wurde 1984 in Berkeley, Kalifornien, von den Designern und Typografen Rudy VanderLans → S. 72 und Zuzana Licko gegründet.

Schriftvertrieb zu gründen. Die Möglichkeit, meine Fonts direkt über meine Website zu verkaufen, bedeutet für mich ein kleines Stück Unabhängigkeit. Die meisten Fonts werden heute von großen Firmen vertrieben. Als Mini-Schriftvertrieb ist es eine Herausforderung, Kunden zu akquirieren, besonders wenn das Budget für die Werbung schmal ist.

Ihre Fonts haben immer einen erweiterten lateinischen Zeichensatz, der sich für eine große Anzahl von Sprachen eignet. Warum?

Mit der OpenType-Technologie sind die Zeichensätze viel größer geworden. Im Laufe der Zeit haben sich die vielen akzentuierten Glyphen zum Standard entwickelt, und es ist ja auch sinnvoll, andere Kulturen miteinzubeziehen, indem man ihre Schriftzeichen mitberücksichtigt. Schriftsysteme sind ein Mikrokosmos mit vielen individuellen Schriftformen und Glyphen-Systemen, die sich fließend parallel zur Sprache entwickeln. Buchstabenformen sind archetypische Zeichen, die ein kulturelles Erbe sind und zugleich die Diversität verdeutlichen. Es ist wichtig, idiosynkratische Sprachen und typografische Gepflogenheiten zu bewahren, die Sprache sichtbar machen, auch außerhalb des lateinischen Alphabets. Viele nicht-lateinische Schriften sind in Gefahr, von dem gebräuchlichsten System verdrängt zu werden, das im Umlauf ist. Mobile Geräte, die typischerweise mit dem lateinischen Alphabet und der breiten Akzeptanz von Englisch arbeiten, fördern diese Entwicklung noch. Schriftdesigner haben eine kulturelle Aufgabe, die verschiedenen Schriftsysteme zu bewahren.

Wenn man Schriftliebhaber oder Studenten nach Schriftdesignern fragt, dann werden hauptsächlich Männer genannt. Bei Fachkonferenzen gibt es mehr Sprecher als Sprecherinnen. Da Sie eine von diesen sind, was kann man Ihrer Meinung nach tun, um mehr Frauen für die Sache zu begeistern?

Völlig unabhängig vom Beruf ist es doch so: Je höher in der Seniorität, desto weniger Frauen findet man. Einer der Hauptgründe ist, dass Frauen öfter als Männer aus dem Beruf aussteigen, zum Beispiel um eine Familie zu gründen. Und wenn sie sich zwischen Beruf und Familie entscheiden müssen, dann beweist das, dass grundlegende Voraussetzungen für Frauen und Familie noch nicht geschaffen sind. Welches Modell genau dazu beitragen könnte, mehr Frauen in Spitzenpositionen zu bringen, ist schwer zu sagen. Eines ist jedoch klar, ob männlich oder weiblich, man braucht eine starke Motivation und individuellen Einsatz, um erfolgreich mitzumischen.

Was ist das Wichtigste, was Sie Ihren Studenten vermitteln?

Dass sie nie vergessen dürfen zu spielen und zu experimentieren. Im Spiel liegt der Ursprung für alles Neue.

Das Interview mit Sibylle Hagmann führten Tabea Dölker und Julia Bielefeld.

Sibylle Hagmann[1] (*1965) grew up in Switzerland and graduated at the Basel School of Design in 1989. In 1996 she went to USA to obtain her Master Degree at the California Institute of Arts, where she discovered her passion for Type and Type Design. Her typefaces won many prizes, e.g. for the type family Odile[2] she received the Swiss Federal Design Award. Hagmann has taught at several schools in California and at the moment she teaches at the University of Houston. She writes for »Sage Publications«, for the »Typographische Monatsblätter« and »IDEA Magazine«. She lives in Houston, where in 2000 she established »Kontour«,[3] her own studio from where she has also been selling fonts since 2012.

Why did you decide to take the Master program in the United States? What made you stay there?

sh: It seemed like a good idea for several reasons. First, I was looking for an entirely different perspective on graphic design and typography, and secondly I was interested in getting work experience abroad. Offers to teach made me stay on in the US, and, having lived there for a couple of years I got used to lots of sunshine.

Was it part of your Master program to get involved with type design?

Not directly, at least it wasn't advertised that way. At that time there were few, if any MFA programs that offered a Masters degree in type design. Jeffery Keedy,[4] (Mr. Keedy as he named himself) the designer of Keedy Sans,[5] an often-discussed typeface at the beginning of the 90s, was a faculty member in the graphic design program at CalArts (California Institute of the Arts, Valencia, California). My biggest hopes were that I would learn from him through the graphic design Masters program how to design and develop digital type. Luckily this came true when he offered a course on the subject during my second year at CalArts.

What is your process of designing a new typeface? Let's take Axia for example. Did it start with the form principle of one specific letter from which the other letters are derived?

The roots of Axia go back to the TwinCities alphabets that I designed for a competition by the Design Institute of the University of Minnesota in 2003. The institute invited several type designers to ponder the question: How can a typeface express what is unique about a city? In the process I experimented with letterforms that followed the conceptual idea of creating a spatial and perspective looking contemporary typeface, formally influenced by local architectural vernaculars. This typeface series had been dormant for a long time, but I knew that I wasn't quite finished with it.

When I started revising the project I had just been commissioned by the Rice School of Architecture to design the school's new identity and logotype in 2011. The logotype I developed employed letters from this typeface I had in the works. The school became increasingly interested to also use the fonts for their print and web designs. With several rounds of revisions, the RSA font family became the predecessor of Axia. With Axia I further assumed the idea of designing a font family that dealt with the constraint of a homogeneous matching text length independent of the weight. Text set in either Light or Black takes up an equal amount of space. Within this setting I experimented with inner arched strokes off the stem on the lowercases n or d, for example, that progressively open the letterforms. That, for example, could be considered a characteristic form concept that has been applied to a series of other lower case letters, and became a defining feature of this typeface.

So, you designed RSA, the predecessor of Axia for the Rice School of Architecture. Is there always a client behind your type design?

Sometimes, but more often there isn't, and I don't mind self-initiating type design projects at all. I enjoy the process very much, and I am appreciative of the fact that the final product is seldom that of many opinions. Often too many cooks take part in the cooking of graphic design solutions. The result is a democratic one, but that doesn't mean it's the best possible result. It works differently with type design. A client or a collaborator is involved in a less direct way.

How big is the difference between your first sketches of a typeface and the look of the final one?

It depends on the typeface and the conceptual ideas behind it. Generally, changes in type design are very often minute; yet they can make a major visual difference. Perhaps the more complex a typeface's concept is, the more differences are there between the sketches and the final version.

How do you know when a typeface is finished?

A typeface family is done when I'm content with every detail, including the formal and technical aspects. Sometimes I think I'm done with a typeface only to rethink it again and with that possibly an entire family in tow. But generally I consider it finished once there isn't an inkling

1 kontour.com

2 Odile is a lively text typeface by Sibylle Hagmann. It was inspired by William A. Dwiggins' type experiments in 1937. In 2006 Odile won the Swiss Federal Design Award.

3 Kontour is Sibylle Hagmann's Studio in Houston/Texas, launched in 2000 and focusing on type design.

4 Jeffrey Keedy (*1957) is an American graphic artist and type designer, whose essays have appeared in Eye, I.D. and Emigre, among others. He is well-known for his typeface Keedy Sans.

5 Keedy Sans is an experimetal sans-serif typeface designed by Jeffrey Keedy and sold through Emigre Fonts.

of doubt about any details. Waking up and not having a long list of things to do with a typeface may be a sign that it's close to being done. I'm usually not rushing on type releases, since a type design is something that matures over time. Over the years I came to rely more on my inner perception and objectiveness to judge when a typeface is finished.

Cholla[6] came out in 1999 and is distributed by Emigre.[7] Why did you decide to found your own label »Kontour«?

Once I decided to focus more specifically on type design, the step to establish my own foundry at the end of 2012 was a logical move. With the selling of fonts from my website I gained some independence. The majority of font sales go through large resellers these days. As a micro foundry, it is very tricky to attract customers, especially when the advertising budget is very modest.

Your fonts always come with an extended Latin character set and cover a large number of languages. Why?

With the arrival of OpenType technology the character sets have become much bigger. Over the years inclusion of many accented glyphs became standard, covering a substantial range of languages is considerate of other cultures. Writing systems are a microcosm consisting of individual letterforms and glyph systems, which develop in flux and in parallel to spoken languages. Letterforms are archetypical signs that constitute a cultural heritage and diversity. It is crucial to preserve idiosyncratic spoken languages and typographic standards that convey visualized language beyond the Latin alphabet. Many non-Latin scripts are in peril of being replaced by the most common system in circulation. Mobile devices, typically equipped with the Latin alphabet and the broad acceptance of English as the lingua franca, further assist this occurrence. Type designers have a cultural responsibility to preserve different writing systems.

When you ask typophiles or students for type designers, mostly male names come up. On type conferences there are more male speakers than female ones. Since you're one of them, what does it need to bring more women into it?

Almost independent of any specific profession, the higher up in the seniority level, the less gender balanced the landscape becomes. One of the main issues is that women tend to jump off a career path at some point due to, for example, raising a family. Having to make a choice between profession and family reveals that core issues haven't been resolved. It is difficult to pinpoint exactly what it would take to get more women involved at the upper end of the ladder. Male or female, finding the motivation to be on the map constitutes an individual effort.

What is the most important thing that you teach your students?

Never forget to play and experiment. Play is the source for anything new that comes about.

Sibylle Hagmann was interviewed by Tabea Dölker und Julia Bielefeld.

6 The type family Cholla (1998–1999) was originally developed for the Art Center College of Design in Pasadena. It won the Award for Excellence in Type Design of the Type Directors Club, New York.

7 Emigre Inc. is an American graphic design studio and was one of the first independent type and software publishers. Emigre was established in 1984 in Berkeley, California, by the designers and typographers Rudy VanderLans → S. 76 and Zuzana Licko.

Diese Zeitung ist ein Organ der Niedertracht. Es ist falsch, sie zu lesen. Jemand, der zu dieser Zeitung beiträgt, ist gesellschaftlich absolut inakzeptabel. Es wäre verfehlt, zu einem ihrer Redakteure freundlich oder auch nur höflich zu sein. Man muss so unfreundlich zu ihnen sein, wie es das Gesetz gerade noch zulässt. Es sind schlechte Menschen, die Falsches tun.

Brandon Text Regular – 7pt

Journalismus im Internet ist nichts anderes als eine Dauerkonversation aller Beteiligten untereinander. Das gedruckte Medium offeriert Geschichten, die aus einem vielschichtigen Diskurs- und Produktionsprozess hervorgehen.

Brandon Grotesque Light – 10pt

Ein Mensch, der gar nichts liest, ist besser informiert als derjenige, der nur Zeitung liest.

Brandon Grotesque Black – 22pt

Unsere Mythologie lesen wir täglich dreimal in der Zeitung.

Brandon Text Bold – 36pt

Bag Grotesque **Bag** Text

Grotesque Thin, Italic	Text Thin, Italic
Grotesque Light, Italic	Text Light, Italic
Grotesque Regular, Italic	Text Regular, Italic
Grotesque Medium, Italic	Text Medium, Italic
Grotesque Bold, Italic	**Text Bold, Italic**
Grotesque Black, Italic	**Text Black, Italic**

Schrift. Typeface.	Brandon
Gestalter. Designer.	Hannes von Döhren
Label. Foundry.	hvdfonts.com
Jahr. Year.	2010–2012

Brandon Grotesque ist eine Groteskfamilie mit sechs Gewichten und passender Kursivschrift. Beeinflusst von den geometrischen Grotesk-Schriften, die in den Zwanziger- und Dreißigerjahren des neunzehnten Jahrhunderts beliebt waren, basieren auch hier die Fonts auf geometrischen Formen, die zum Zwecke der besseren Lesbarkeit optisch korrigiert wurden. Brandon Grotesque hat ein funktionelles, aber warmes Schriftbild. Während die feinen und die schwarzen Gewichte sich bestens für Displaygrößen eignen, sind die leichten, normalen und mittleren Gewichte für längere Texte geeignet. Die geringe x-Höhe und die zurückhaltenden Formen geben der Schrift eine unaufdringliche Eleganz. Brandon Grotesque ist prädestiniert für anspruchsvolle professionelle Aufgaben.

Brandon Text hat eine größere x-Höhe als die Grotesque Version und ist optimiert für lange Texte, kleinen Druck und Bildschirme. Die gesamte Brandon Serie verfügt über alternative Buchstaben, Brüche und einen erweiterten Zeichensatz für ost- und westeuropäische Sprachen.

Brandon Grotesque is a sans-serif type family of six weights plus matching italics. Influenced by the geometric-style sans-serif faces that were popular during the 1920s and 30s, the fonts are based on geometric forms that have been optically corrected for better legibility. Brandon Grotesque has a functional look with a warm touch. While the thin and the black weights are great performers in display sizes, the light, regular and medium weights are well suited for longer texts. The small x-height and the restrained forms lend it a distinctive elegance. Brandon Grotesque is well equipped for complex, professional typography.

Brandon Text has a higher x-height than the Grotesque version and is optimized for long texts, small sizes and screens. The whole Brandon series is equipped with alternate letters, fractions and an extended character set to support East European as well as West European Languages.

Monatsschrift

KOSMOS

ZEIT IM BILD

»Art Déco«

DER STERN

15. Jahrgang

Brandon Grotesque Medium
Brandon Text Medium

Brandon Grotesque Black Italic
Brandon Text Light Italic

Brandon Text Black
Brandon Grotesque Regular

Der in Berlin geborene *Hannes von Döhren*[1] (*1979) studierte Grafik-Design und arbeitete in einer Hamburger Werbeagentur. 2008 gründete er in Berlin seine Type Foundry »HVD Fonts«[2] und veröffentlichte in der Folge zahlreiche Schriftfamilien wie Brevia, Livory (in Zusammenarbeit mit Livius Dietzel),[3] ITC Chino, FF Basic Gothic, Reklame Script, Pluto und Pluto Sans. Bei »MyFonts« war seine *Brandon Grotesque* die erfolgreichste Schrift des Jahres 2010. 2011 wurde Hannes von Döhren durch den Type Directors Club New York mit dem Certificate of Excellence in Type Design ausgezeichnet. 2013 erschien mit der Brandon Text eine auf Lesegrößen optimierte Variante der Brandon Grotesque.

Du arbeitest in Berlin von Zuhause aus. Wie kann man sich Deine Arbeitsatmosphäre vorstellen?

Ich habe meine Wohnung in einen Wohn- und einen Bürobereich unterteilt. Die Schriften entstehen bei mir zu Hause. Ich arbeite aber auch außerhalb, wenn ich z.B. von Agenturen gebucht bin. Viele Projekte entstehen außerdem über E-Mail-Austausch – ich habe Kunden aus der ganzen Welt.

Du hast eine Zeitlang in Hamburg in einer Werbeagentur gearbeitet. Wie verlief Deine Entwicklung vom Werber zum Type Designer?

Mein Fokus lag im Studium auf Grafik-Design. In der Werbeagentur musste ich mich erst mal umstellen – Werbung hat einen etwas anderen Ansatz als Grafik-Design. Das war eine spannende Zeit, in der ich viel Neues dazu gelernt habe.

Meine Liebe zur Typografie hat mich allerdings immer schon begleitet und so habe ich parallel zu meinem Job in der Agentur angefangen, am Wochenende mit Freunden und Kollegen Schriften zu basteln. Zuerst aus Kartoffeln geschnitzt oder aus Pappe ausgeschnitten. Das war für mich gewissermaßen ein Ausgleich zu der zielorientierten Arbeit für den Kunden.

Ich habe mich dem Thema Schriftentwicklung zuerst eher von der experimentellen Seite genähert. Die ganzen Free Fonts, die es jetzt noch von mir gibt, sind alle in meiner Freizeit in Hamburg entstanden.

Ich war erstaunt, wie viele Leute diese Fonts heruntergeladen und wie sich die Schriften verbreitet haben. Ich wollte mich weiterentwickeln und eine Schrift gestalten, die ich auch verkaufen könnte – seriöser und auch tatsächlich einsetzbar.

Wie unterscheidest Du all' die verschiedenen Schriften?

Mit der Zeit bekommt man ein geschulteres Auge. Am Anfang sucht man sich beim Schriftenidentifizieren immer die Charakterbuchstaben der Schrift. Die Meta[4] erkennt man z.B. immer am offenen g. Irgendwann fängt man an, ein Gefühl für die gesamte Schrift zu bekommen:

1 hvdfonts.com

2 Die Berliner Type Foundry HVD Fonts, die Display- und Textschriften vertreibt, wurde 2005 von Hannes von Döhren gegründet.

3 Livius Dietzel (*1979) studierte Visuelle Kommunikation an der Berliner UdK, in London sowie in Barcelona. Neben seiner gestalterischen Tätigkeit bei MetaDesign Berlin arbeitet er als freier Art Director und Schriftentwerfer.

4 Die humanistische serifenlose Linear-Antiqua FF Meta wurde von Erik Spiekermann gestaltet. Ursprünglich als Auftragsarbeit für die Deutsche Bundespost entworfen, zählt sie heute zu den bekanntesten zeitgenössischen Groteskschriftfamilien. Sie wurde 1991 in der FontFont-Bibliothek veröffentlicht.

Man merkt, eine Akzidenz Grotesk[5] ist nicht dasselbe wie eine Frutiger[6] oder eine Helvetica.[7] Man schaut dann gar nicht mehr nur auf die Charakterbuchstaben, sondern erkennt die Schrift schon am Schriftbild. Wenn bei Kaiser's dann »Schnitzel« steht und da kein g drin ist, erkennt man trotzdem, dass es die Meta ist. Bei mir hat sich bei der Schriftgestaltung mit der Zeit auch das Interesse vom einzelnen Buchstaben zu der Frage verlagert, wie eine Schrift in sich funktioniert. Der nächste Schritt ist dann in einer Schriftfamilie z.B. von Hairline bis Black zu denken. Und den Charakter der Schrift über alle Gewichte sichtbar zu machen.

Wie kam es dann zur Gründung einer eigenen Foundry?

Die Quench[8] war meine erste kommerzielle Schrift. Ich bekam von Linotype[9] eine Zusage und eine Coding-Tabelle mit ca. 400 Zeichen, die ich dann vervollständigen sollte. Nachdem ich den Font fertiggestellt hatte, dachte ich: »Yeah, jetzt geht's los!«. Dann hat es noch fast ein Jahr gedauert, bis der Font endlich auf dem Markt war. Mir war gar nicht bewusst, wie viel Arbeit nach der Fertigstellung der Zeichen noch in eine Schrift investiert werden muss. Zusätzlich hat eine große Foundry eine Menge anderer Schriften im Portfolio, um die sie sich »kümmern« muss.

So begann ich Schriften über MyFonts zu veröffentlichen – im Vergleich zu einer Foundry, die eine Schrift auf Qualität prüft und sich um das Marketing kümmert, ist MyFonts ein reiner Distributor.

Da muss man eben alles selbst machen: Die fertigen Font-Dateien erstellen, Qualitätscheck, Grafiken und Promotion – selbst den Kundenservice, wenn es um die Schriften oder Problemlösungen geht. Dafür hat man selbst das Timing in der Hand. Da ich das alles gerne selbst mache, passte das für mich besser. Aber dieses Wissen hätte ich, ohne den Vergleich je selbst gemacht zu haben, so nicht gehabt. Über die Jahre entstand dann sozusagen Stück für Stück meine eigene kleine Foundry: HVD Fonts.

Woher nimmst Du die Inspiration für Deine Schriften?

Schriften sind oft für bestimmte Zwecke gemacht und gedacht. Einfach gesagt: Die Comic Sans[10] ist eher etwas Lustiges, die Klavika[11] ist eher etwas Technisches. Jede Schrift ist anders und vermittelt auch ein anderes Gefühl. Woran das liegt, ist für den Laien oft gar nicht erkennbar.

Inspiration finde ich in vielen alltäglichen Dingen. Auf Berlins Straßen findet man sie an Hauswänden oder auch auf alten Werbeplakaten. Inspirierend kann nur ein einzelner Buchstabe oder eine abstrakte Form sein.

Ich möchte Schriften entwickeln, die auch in der Anwendung funktionieren. Ich wünsche mir selbst manchmal eine ganz bestimmte Schrift für ein Projekt – wenn es die noch nicht gibt, mache ich sie einfach. Ich habe in meinem Entwurfsblock noch eine ganze Menge Ideen, die ich gern umsetzen würde und es kommen auch immer wieder neue hinzu.

5 Die Akzidenz Grotesk von Günter Gerhard Lange ist eine Sans Serif-Schriftenfamilie. Sie wurde 1896 von der Berthold Type Foundry als Accidenz-Grotesk veröffentlicht.

6 Frutiger ist eine serifenlose Linear-Antiqua-Schrift, die 1975 von Adrian Frutiger entworfen und von der Schriftgießerei D. Stempel veröffentlicht wurde (→ S. 82).

7 Die Schriftfamilie Helvetica ist eine serifenlose Linear-Antiqua mit klassizistischem Charakter und gehört zu den am weitesten verbreiteten Groteskschriften. Die ersten Schriftschnitte wurden ab 1956 von Max Miedinger und Eduard Hoffmann gestaltet (→ S. 82, 183).

8 Quench ist eine von Hannes von Döhren gestaltete und im Jahr 2008 bei Linotype veröffentlichte Schriftfamilie. Die winkligen Ecken der Punzen stehen im Kontrast zu den abgerundeten Außenformen.

9 Die 1890 gegründete Monotype GmbH vertreibt bzw. lizensiert heute hochwertige digitale Schriften über monotype.com Außerdem entwickelt Monotype Corporate Fonts für Geschäftskunden und stellte 2005 das Font Management-Programm Font Explorer zur Verfügung (→ S. 60, 80).

10 Vincent Connare entwickelte die Comic Sans 1994 während seiner Tätigkeit für Microsoft. Die handschriftähnliche Sans Serif-Schrift sollte ursprünglich in den Sprechblasen der Software Microsoft Bob eingesetzt werden, die zur Benutzerführung gedacht waren.

11 Die 2004 von Eric Olson gestaltete Klavika wurde bei der Process Type Foundry veröffentlicht und ist besonders durch die Verwendung im Facebook-Logo populär geworden.

HANNES VON DÖHREN

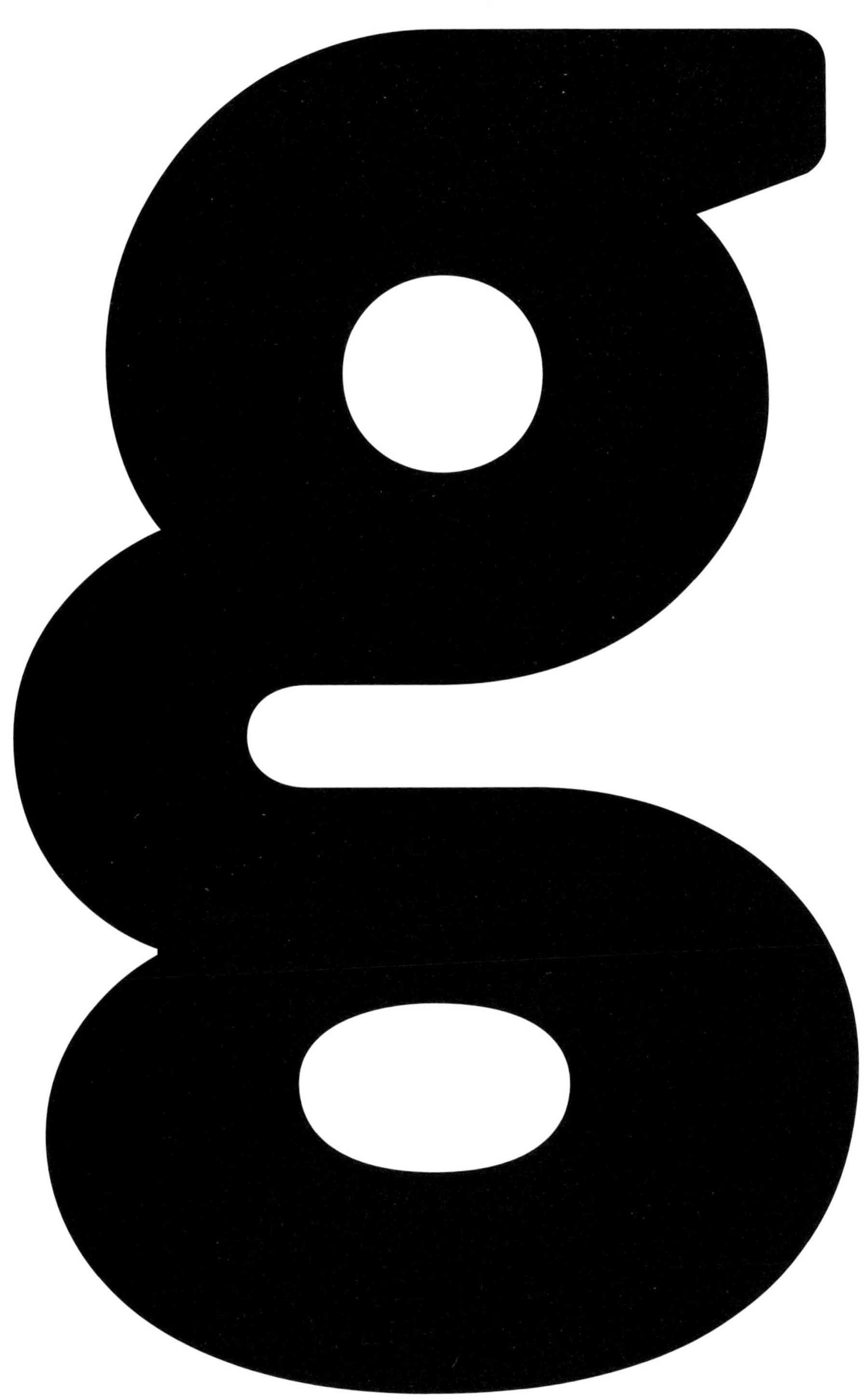

Die Brandon Grotesque war ja sehr erfolgreich. Kannst Du sie kurz in eigenen Worten beschreiben? Wofür steht sie?

Ich war fasziniert von der Haptik und der intensiven Aura, die Magazine aus den 1920er und 1930er Jahren ausstrahlten – wie die Fließtexte gesetzt wurden und welche Schriftmischungen verwendet wurden! Ich wollte unbedingt eine Schrift mit dieser Anmutung gestalten. Eine geometrische Schrift, die dennoch eine gewisse Weichheit und Wärme besitzt. Durch den »schlechten« Druck erschienen die Ecken der Textschriften leicht abgerundet – ein Gefühl, das heute, in der Zeit der gestochen scharfen Druckbilder, nicht mehr aufkommen kann.

Ich entschied mich, die Brandon mit leicht abgerundeten Ecken zu versehen – sie sollte trotz ihrer Geometrie und Klarheit etwas Wärme ausstrahlen. Obwohl die Brandon Grotesque mit 12 Schnitten eine relativ große Familie ist, hat jeder Schnitt seine eigene Ästhetik: Die Schnitte basieren zwar aufeinander, wurden aber alle einzeln gezeichnet. So konnte ich jeden Schnitt mit eigenen Details ausstatten.

Du hast ja jetzt noch eine Textversion mit größerer x-Höhe herausgebracht. War das von Anfang an geplant?

Nein, ich hatte gar nicht erwartet, dass die Brandon Grotesque so erfolgreich wird. Ich finde sie perfekt, so wie sie ist, für sehr kleine Größen ist sie aber nicht optimal, da sie durch die niedrige x-Höhe in diesen Größen noch kleiner wirkt. Ich habe sehr viele Anfragen von Gestaltern bekommen, eine Variante für kleine Größen zu machen.

Ich hatte schließlich selbst einen Auftrag, bei dem ich mit der Brandon gearbeitet und dann bemerkt habe, dass ich gerne eine Text-Variante hätte. Also habe ich sie gezeichnet und für den Einsatz in Lesegrößen und am Screen optimiert. Ich bin sehr froh darüber, dass jetzt ein viel breiteres Feld mit der Brandon-Familie abgedeckt werden kann.

Gibt es abgesehen von der x-Höhe noch andere Details, die die Text von der Grotesque abheben und sie besonders machen?

Ja, das kleine g der Brandon Grotesque ist z.B. sehr speziell – wie eine Schlange. Ich mag es auch richtig gern. Es ist ein richtiger Charakterbuchstabe. Wenn ich es aber im Fließtext sehe, habe ich immer das Gefühl, dass diese Form zu komplex ist. Der Grauwert ist dann nicht mehr so gleichmäßig. Nach einigen Tests und unzähligen Ausdrucken habe ich mich schweren Herzens bei der Brandon Text von dem zweistöckigen g getrennt und es durch ein simples, klares g ersetzt. So ist der Grauwert wesentlich gleichmäßiger. Ich habe bei der Brandon Text auch diverse Buchstabenbreiten optimiert, um den Leserhythmus zu verbessern.

In letzter Zeit sind sehr viele kleine Foundries dazu übergegangen, kostenlose Testfonts mit beschränktem Zeichensatz anzubieten. Denkst Du, dass sich dieses Modell durchsetzen könnte?

Es ist auf jeden Fall gut, wenn die Kunden die Schrift vorher testen können und direkt sehen können, ob sie passt oder nicht. Bei einigen meiner Schriften biete ich auch einzelne Schnitte als kostenlose Demo-Fonts an.

Der Nachteil ist vielleicht, dass einige Leute die kompletten Schriften nicht mehr kaufen, wenn ihnen der reduzierte Zeichensatz der Testfonts ausreicht. Wenn man wie ich mit Schriftverkauf seinen Lebensunterhalt finanzieren muss, wird man dieses »Risiko« abwägen müssen. Ich selbst konnte mich auch noch nicht zu einer klaren Entscheidung durchringen.

Ich würde mir als Grafik-Designer wünschen, alle Schriften testen zu können, würde dann aber auch im Gegenzug die Vollversionen bei Verwendung kaufen. Ich glaube es ist wichtig, ein Bewusstsein bei den Nutzern zu schaffen, dass da auch Menschen hinter den Schriften stehen und da eine Menge Arbeit drin steckt. Wer eine Schrift kauft, gibt die Unterstützung, aber auch ganz wichtig: den Respekt für die Arbeit des Schriftentwerfers.

Das Interview mit Hannes von Döhren führten Yvonne Kümmel und Jens Giesel.

Hannes von Döhren[1] (*1979), who was born in Berlin, studied graphic design and worked at an advertising agency in Hamburg. In 2008, he founded his »HVD Fonts«[2] type foundry in Berlin. In the following time he published numerous font families like, for instance, Brevia, Livory (in collaboration with Livius Dietzel),[3] ITC Chino, FF Basic Gothic, Reklame Script, Pluto und Pluto Sans. His _Brandon Grotesque_ was the most successful font of the year 2010 at »MyFonts«. In 2011, Hannes von Döhren was awarded the Certificate of Excellence in Type Design by the Type Directors Club New York. In 2013, the Brandon Text font was published. It was a variant of the Brandon Grotesque font (optimized with regard to reading sizes).

You are working in Berlin from home. How are we to imagine your workplace?

I have divided my apartment into a living and a working area. The typefaces are made in my home, but I also work outside, e.g. if I am booked by an agency. Many projects, however, are carried out via e-mail – I have customers from all over the world.

You have been working for an advertising agency in Hamburg for a while. How did your development from advertising man to type designer come about?

During my studies, my emphasis was on graphic design. In the advertising agency I had to re-think, because here the approach is somewhat different from graphic design. That was an exciting time, where I learned a lot.

However, I have always loved typography, and so I started to experiment with typefaces, usually at weekends, together with some friends and colleagues. At first we carved them from potatoes or cut them from cardboard. For me this was a welcome change from the very goal-oriented commission work. Initially, my approach to type design was rather experimental. All those Free Fonts, which still exist, were created in my spare time in Hamburg.

I was surprised at how many people downloaded those fonts and how they spread. But I wanted to develop it and design a typeface I could sell – more serious and truly useful.

How do you differentiate between all those typefaces?

Over time you develop an eye for it. At first you identify a typeface by its characteristic letters. The Meta,[4] for instance, you can always recognize by the open g. Then gradually, you start to develop a feeling for the typeface as a whole: you realise that an Akzidenz Grotesk[5] is not the same as a Frutiger[6] or a Helvetica.[7] You don't look at just the characteristic letters any more, you recognize the typeface. And if you see the word »Schnitzel« at the supermarket, you see that it is the Meta, even if there is no g in the word. For me, when designing a new typeface, the interest in the single letter has more and more given way to the question of how a typeface works as a whole. The next step then is to imagine a whole type family, e.g. from Hairline to Black, and to maintain the characteristics of a typeface across all weights.

How did the launching of your own foundry come about?

The Quench[8] was my first commercial typeface. I got an agreement from Linotype,[9] together with a coding table of about 400 characters which I was asked to complete. When I had finished with this font, I thought, »Yeah – that's it!« But then it took another year before the font was ready for publication. I had no idea how much work there is still required after the characters of a typeface are finished. And of course a large foundry has lots of other typefaces in its portfolio which also need looking after.

So I started to publish typefaces via MyFonts. In comparison with a foundry which checks a typeface for quality and sees to the marketing, MyFonts is purely a distributor. You have to do everything yourself: creating the completed font data, quality control, graphics and promotion – even customer service, if it is a question of types or problem solving. On the other hand, you are in charge of the timing. And it suited me, because I like doing all those things. But I wouldn't have known what is involved if I hadn't had the chance to learn it for myself. And so, over the years, my own small foundry took off: HVD Fonts.

1 hvdfonts.com

2 The Berlin type foundry HVD Fonts, selling display and text fonts, was launched in 2005 by Hannes von Döhren.

3 Livius Dietzel (*1979) studied Visual Communication at the Berlin UdK, as well as in London and Barcelona. Apart from working for Meta Design, Berlin, he is a freelance art director and type designer.

4 The sans-serif Linear-Antiqua FF Meta was developed by Erik Spiekermann. Originally designed for the Deutsche Bundespost, it is now one of our best-known contemporary grotesque typefaces. Is was published in 1991 in the FontFont Library.

5 The Akzidenz Grotesk by Günter Gerhard Lange is a sans-serif type family. It was published in 1896 by the Berthold Type Foundry as Accidenz-Grotesk.

6 Frutiger is a sans-serif Linear-Antiqua type, designed by Adrian Frutiger in 1975 and published by the type foundry D. Stempel (→ S. 87).

7 The type family Helvetica is a sans-serif Linear-Antiqua with classicistic character and is one of the most frequently used grotesque typefaces. The first type styles were designed from 1956 by Max Miedinger and Eduard Hoffmann (→ S. 87, 185).

8 Quench is a type family designed in 2008 by Hannes von Döhren. The angular interletter spaces form a contrast to the rounded outer forms.

9 Monotype GmbH, established in 1890, sells and licences high quality digital fonts via monotype.com. Monotype also develops corporate fonts for clients and in 2005 made the Font Management Programme Font Explorer available (→ S. 63, 85).

Where do you get the inspiration for your typefaces?

Typefaces are often created for a certain purpose. Simplified: the Comic Sans[10] is rather playful and amusing, the Klavika[11] is more technical. Each typeface is different and creates a different feeling, and the layman often doesn't understand why this happens.

I get my inspiration from many everyday things or situations. In Berlin you can find it on house walls or on old posters, and you can get inspired by just a single letter or by an abstract shape.

I want to develop typefaces that work in practical use. Sometimes I am looking for a very definite typeface for a certain project – and if it doesn't already exist, I will make it. I still have many ideas on my sketch pad which I want to realise, and I keep adding new ones.

The Brandon Grotesque was highly successful. Could you briefly describe it in your own words? What does it stand for?

I was fascinated with magazines from the Twenties and Thirties, which radiated an intense aura and tactile qualities – the setting of the continuous texts and the mixture of typefaces – and I badly wanted to create a typeface with this feeling. A geometric sans-serif, but still with a certain softness and warmth. The imperfect print in those days made the corners appear slightly rounded – an impression we don't get any more in this day and age of needle-sharp print.

I decided to give the Brandon slightly rounded corners – I wanted it to have some warmth, in spite of being clear and geometric. And although the Brandon Grotesque is a comparatively large family with its 12 styles, each style has its own aesthetics: they are based upon each other, but each of them was drawn individually. In this way I could give each style its own details.

You have now come up with a version showing a greater x-height. Did you plan that right from the start?

No, I hadn't expected the Brandon Grotesque to be such a success. I think it is perfect as it is, except for very small sizes, because there the small x-height makes it look even smaller. I've had many requests from designers to add a variant for small point sizes. Finally I had a commission myself where I worked with the Brandon and noticed that I would like a text variant. So I drew it and then optimized it for reading and screen work. I am very glad that the Brandon family now covers a much wider range.

Apart from the x-height, are there any other details in which the Text differs from the Grotesque and makes it special?

Yes, for instance the small g in the Brandon Grotesque is very special – like a snake. I really like it, it is one of those characteristic letters. However, if I see it in a continuous text I always think that its form is too complex. The colour isn't so even any more. After much experimenting and countless print-outs I finally and reluctantly decided to do away with the two-storey g and replaced it by a clear and simple g. Now the colour is much more even. In the Brandon Text I also optimized a few character widths to improve the reading rhythm.

Many small foundries have lately started to offer free test fonts with a limited set of characters. Do you think that this approach will catch on?

I do think it is a good thing if customers can first test a typeface to see whether it meets with their requirements. For some of my typefaces I also offer certain styles as free trial fonts.

It may be a disadvantage that then some people will not buy the complete family, because the limited character set is sufficient for them. If you have to earn your living by selling typefaces, as I do, you have to weigh up those risks very carefully. So far, I haven't yet reached a final decision about that.

As a graphic designer I would wish to be able to test all typefaces, but then, if I am going to use them, I would buy the complete set. I think it is important to create an awareness with the users that there are people behind those typefaces, and that they have invested a lot of time and trouble. A customer buying a typeface supports this, but just as important: he respects the work of the type designer.

Hannes von Döhren was interviewed by Yvonne Kümmel and Jens Giesel.

10 Vincent Connare developed the Comic Sans in 1994 while working for Microsoft. The sans-serif typeface, which resembles handwriting, was originally supposed to be used for the speech bubbles of the Microsoft Bob software which was developed for user guidance.

11 The Klavika, designed by Eric Olson in 2004, was published by the Process Type Foundry. Is has achieved a wide popularity through its use in the Facebook logo.

Das Amphitheatrum Flavium (erst später umbenannt in Kolosseum) ist das *größte antike Amphitheater der Welt,* Ort von Brot und Spielen und Gladiatorenkämpfen. Zeugnis für Baukunst und grausame Spektakel.

Im Jahre 72 nach Christus wurde der Bau des Kolosseums in Rom durch Vespasian begonnen, um 80 wurde es der Überlieferung nach mit *100 Tage andauernden Spielen* eröffnet, u.a. mit Gladiatoren kämpfen, nachgestellten Seeschlachten und Tierhetzen.

Capitolium News Regular, Italic – 7pt

Als Zuschauer gelangte man durch eines der *80 Tore* ins Innere der Arena. Einige Eingänge und Bereiche des Zuschauerraums waren für privilegierte Gäste reserviert. Im Inneren war Platz für *50.000 Zuschauer.* Viele damalige Prinzipien der Gestaltung der Arena werden auch heute noch beim Bau von Stadien eingesetzt.

Capitolium News Medium, Medium Italic – 10pt

Auf einer ellipsenförmigen Grundfläche erbaut, hat das Kolosseum einen Umfang von 527 m. Die runde Form wurde gewählt, dass Gladiatoren und gehetzte Tiere keine Ecke hatten, in der sie hätten Schutz suchen können.

Capitolium News SemiBold Italic – 18pt

Nach Entfernung der Holzdielen konnte der Boden der Arena geflutet werden und Seeschlachten wurden nachgestellt.

Capitolium News Bold – 24pt

Schrift. Typeface. Capitolium News
Gestalter. Designer. Gerard Unger
Label. Foundry. gerardunger.com
Jahr. Year. 2006

Capitolium wurde 1998 für das Jubiläum der römisch-katholischen Kirche im Jahre 2000 entwickelt. Dieser Schriftentwurf war der zentrale Teil eines Projekts zur Gestaltung eines Informations- und Leitsystems durch Rom für Pilger und Touristen. Mit Capitolium wird auch die fast ununterbrochene zweitausend Jahre alte Tradition der öffentlichen Beschriftung in der Stadt Rom fortgesetzt. Es ist eine moderne Schrift für das 21. Jahrhundert, das sich jedoch stark an der römischen Tradition orientiert.

Obwohl Capitolium sich gut für die meisten modernen Produktionsprozesse und auch für den Bildschirm eignet, ist die Schrift für Zeitungsdruck zu fragil. Dafür bedurfte es einer robusteren Schrift mit mehr Zeichen pro Textzeile. Capitolium News verfügt über eine größere x-Höhe als Capitolium. Capitolium News hat ein modernes Schriftbild, deren klassische Formen einer großen Tradition verbunden sind.

Capitolium News für fortlaufende Texte gibt es in den Variationen Normal, Kursiv, Halbfett, Halbfett kursiv, Fett und Fett kursiv, für Überschriften gibt es Leicht, Normal, Halbfett und Fett.

Capitolium was designed in 1998 for the Jubilee of the Roman Catholic Church in 2000. This type design was the central part of the project for a guiding and information system for pilgrims and tourists to find their way through Rome. Capitolium also continues Rome's almost uninterrupted two-thousand-year-old tradition of public lettering. It is a modern typeface for the twenty-first century and strongly related to the traditions of Rome.

Though Capitolium works well in most modern production processes and also on screens, it is too fragile for newsprint. For newspapers, sturdier shapes were required as well as more characters to a line of text, and Capitolium News has a bigger x-height than Capitolium. Capitolium News is a thoroughly modern newsface, with classic letterforms linked to a strong tradition.

Capitolium News for running text comes in the variations Regular, Italic, Semibold, Semibold Italic, Bold and Bold Italic, and for headlines as Light, Regular, Semibold and Bold.

BRØDRE

Rómulo

Româna

753 B.C.

Faustulus

Capitolium News Medium
Capitolium News Bold Italic
Capitolium News Medium
Capitolium News Italic
Capitolium News SemiBold

Gerard Unger[1] (*1942) studierte bis 1967 an der Gerrit Rietveld Akademie und begann 1972 als freiberuflicher Designer und Schriftgestalter zu arbeiten. Über dreißig Jahre lehrte er an der Rietveld Academie; derzeit bekleidet Unger Professuren in Reading[2] (GB) und Leiden (NL). Seine Schriften und Informationsschriften wurden mehrfach international ausgezeichnet, unter anderem 1984 mit dem H. N. Werkman-Preis und dem Piet Zwart-Preis 2012. Ungers Publikation »Terwijl je leest«[3] (Dt.: »Wie man's liest«), die sich mit dem Prozess des Lesens beschäftigt, ist in zahlreiche Sprachen, zuletzt Koreanisch, übersetzt worden.

Wie kamen Sie zum Type Design?

gu: Ich war kein besonders guter Schüler bis zu dem Moment, an dem ich an die Gerrit Rietveld Akademie kam. Ich habe immer Schwierigkeiten mit dem Lesen gehabt. Ein Kollege von mir hat meine Berufswahl als Kompensation bezeichnet. Während meiner Grundschulzeit und am Gymnasium lief es nicht gut. Auf der Rietveld habe ich dann plötzlich entdeckt, dass ich wirklich etwas kann. Dieser Moment der Entdeckung, dass ich der Welt selbst etwas hinzufügen kann, hat mir so viel Spaß gemacht und das tut es heute immer noch, so dass ich mich jeden Tag an meinen Computer setze.

Wie sieht der Prozess des Schriftgestaltens bei Ihnen aus? Welche Arbeitsmaterialien benutzen Sie? Wie beginnen Sie?

Seit 1986, als ich meinen ersten Mac gekauft habe, habe ich nicht mehr mit der Hand gezeichnet. Zuvor arbeitete ich am Leuchttisch mit Transparentpapier und zog feine Bleistiftlinien. Als ich dann angefangen habe auf dem Mac zu arbeiten, bemerkte ich schnell, dass es ein viel direkteres Arbeiten ist. Man hat den Computer, man hat den Drucker und man gestaltet Formen. Man kann einen Direktdruck machen. Dann schaut man ihn an, merkt, was verändert werden muss und arbeitet weiter. Ab und zu skizziere ich noch als Zwischenphase, wenn ich nicht weiterkomme und im Kopf – nicht am Computer – am Ende der Möglichkeiten bin. Der Computer lässt einfach alles zu. Ich kann z.B. alle meine Kurven ändern oder die Serifen weglassen. Das kann man alles sehr schnell machen und viel Freizeit damit verbringen.

Ihre Schriften werden für Zeitungen, Magazine, Straßenschilder und Bücher verwendet. Wissen Sie schon von Anfang an, was der Gestaltungszweck sein wird?

Ich kann mir vornehmen, eine Zeitungsschrift zu gestalten. Aber dann kommen die Benutzer und sagen: »Das ist eine wunderbare Schrift für ein Buch.« Andere sagen: »Das ist eine wunderbare CI-Schrift!« Meine Studenten kommen zu mir: »Ich will eine Schrift speziell für eine Kinderenzyklopädie machen.« Ich sage dann: »Okay, mach mal. Aber rechne damit, dass die Benutzer sie für völlig andere Zwecke einsetzen.« Denn da ist ein Raum zwischen Gestalter und

1 gerardunger.com

2 Die University of Reading ist eine Universität im englischen Reading. Sie wurde 1926 gegründet. Besonders einflussreich ist der Masterstudiengang Typeface-Design. Viele der in diesem Buch genannten Schriftgestalter haben dort studiert bzw. gelehrt (→ S. 178).

3 Gerard Ungers »Terwijl je leest« (dt.: »Wie man's liest«) beschäftigt sich mit dem Prozess des Lesens. Die Publikation wurde bereits in zahlreiche Sprachen übersetzt. Vgl. Unger, Gerard: Wie man's liest. Sulgen: Niggli, 2009.

Benutzer. Eigentlich habe ich im Hinterkopf immer die Idee, dass es nicht nur eine zweckmäßige Schrift sein darf. Sie muss auch einen breiten Anwendungsraum haben. Das ist einfach die Praxis.

Sie greifen auch mal zum Pinsel, experimentieren. Wie viel Zufall steckt in Ihren Schriften?

Da ist immer eine Art von Zufall am Werk. Es gibt Momente in der Gestaltung, in denen plötzlich alles gut geht – von alleine. Die Arbeit ist fertig, man ist erstaunt und fragt sich, ob man es selbst gemacht hat. In diesen Momenten steckt einfach viel Erfahrung, die mitwirkt und man wird mitgenommen von Ideen, Erfahrung, Vision. Das ist mir z.B. mit der Swift[4] passiert, da wusste ich am Anfang, was ich machen möchte. Aber dann gab es einen Moment, als ich beim Skizzieren plötzlich bemerkte, dass etwas passiert und ich habe rasch weitergearbeitet – fast 24 Stunden am Stück, ohne Schlaf und Essen. Aber es hat auch Schriften gegeben, bei denen man sich selbst richtig zwingen muss. Manchmal muss man die Ergebnisse einen Monat in die Schublade legen und kann erst dann wieder weitermachen.

Wie wichtig ist Ihnen Zusammenarbeit bei der Schriftgestaltung?

Wenn man will, ist Schriftgestaltung etwas sehr Einsames. Aber auch wenn ich Schriften in eigenem Auftrag gestaltet habe, wie die Swift, habe ich immer Rücksprache gehalten mit Menschen aus der Praxis, z.B. Zeitungsgestaltern. Ich habe sie gefragt, was ihre Wünsche sind, wie sie über dieses oder jenes denken. Ich habe immer versucht, als Schriftgestalter so sozial wie möglich zu sein. Wenn man stundenlang hinter dem Computer sitzt – das ist Einsamkeit. Und das ertrage ich nicht zu lange.

Für die letzte Schrift, die ein sehr großes Character-Set von fast mehreren tausend Zeichen hat, mit polyphonischem Griechisch dabei, zukünftig vielleicht noch Koreanisch, Japanisch, Arabisch, Devanagari, hatte ich viele ehemalige Studenten, mit denen ich zusammengearbeitet habe. Auch in der Produktion der Schrift arbeite ich mit ehemaligen Studenten wie Tom Grace aus Heidelberg oder Irene Vlachou, beide ehemalige Studierende aus Reading. Ich mache das Basisset von Zeichen und sie arbeiten diese aus. Er macht die Kapitälchen, die Mediävalziffern etc. und zusammengestellte Zeichen für viele Sprachen, sie hilft mir mit Griechisch. Überarbeitet kommen die Zeichen zurück und ich kann sagen: »Nein, diese Kurve muss ich neu überarbeiten, damit das wieder meine Kurve wird.«

Wie kommen Sie auf die Namen Ihrer Schriften, wie z.B. Gulliver[5] oder Swift?

Das ist das Schwierigste. Mein Kollege Matthew Carter[6] hat gesagt: »Ich gestalte lieber zehn neue Schriften, als dass ich mir den Namen für eine ausdenken muss.« Bei mehreren Schriften wusste ich den Namen schon bevor ich die Schrift gestaltet habe, wie bei der Swift. Das ist Englisch für »Mauersegler« und es hat mir als Inspiration gedient, dass die Vögel immer in der Luft über mir waren und mich angespornt haben. Bei der Gulliver kam es etwas später. Durch Experimente habe ich realisiert, dass ich am Gestalten einer Schrift war, die in den ganz kleinen Graden viel größer aussieht, als sie wirklich ist und die man in viel kleineren Graden benutzen konnte als viele andere Schriften. Da wusste ich, das ist »Gulliver«. Bei den Liliputanern war er ein Riese und unter den Riesen war er ein Zwerg. Ein wunderbar passender Name. Aber es gab auch Schriften, bei denen es lange gedauert hat.

Sie vertreiben die Capitolium 2[7] über TypeTogether.[8] Wie kam es zu dieser Zusammenarbeit?

Ich habe mehr als zehn Jahre meine Schriften selbst vertrieben. In meinem Atelier kamen die

4 Unger entwickelte die Swift zwischen 1984 und 1987 als eine moderne digitale Schrift für Zeitungen.

5 Die Gulliver ist eine sehr platzsparende und gut lesbare Zeitungsschrift, die Unger 1993 fertiggestellt hat und von der er nur 100 Lizenzen exklusiv vertreibt.

6 Matthew Carter ist ein englischer Schriftgestalter mit über fünfzig Jahren Erfahrung, in denen er für verschiedene Linotype-Gesellschaften in den USA und in Europa gearbeitet hat. 1981 war er Mitbegründer der Schriftenfirma Bitstream Inc., Cambridge (Massachusetts), die er 1991 zusammen mit Cherie Cone verließ, um dann Carter & Cone Type, Inc. zu gründen.

7 Die Capitolium wurde 1998 anlässlich des Jubiläums der römisch-katholischen Kirche als Schilder- u. Informationsschrift für ein Leitsystem für Pilger und Touristen durch Rom entwickelt. Überarbeitet für den Zeitungssatz wurde sie 2006 als Capitolium 2 bzw. Capitolium News herausgegeben.

8 Nachdem sich José Scaglione und Veronika Burian während ihres Master-Studiums an der University of Reading kennengelernt hatten, gründeten sie 2006 die Foundry TypeTogether (→ S. 32).

Bestellungen an und wurden am selben Tag bearbeitet. Am Anfang mit Floppys per Post, später ganz einfach per E-Mail. Es war wieder dieses alte Problem der Einsamkeit: Wenn ich mich den ganzen Tag mit Bestellungen beschäftige und auch noch versuche, meiner eigenen Arbeit als Schriftgestalter, als Grafiker und Typograf nachzugehen, dann muss etwas abgegeben werden. Veronika Burian[9] und José Scaglione[10] sind ehemalige Studierende aus Reading. Und ich habe mir gedacht, wenn man Studenten ausgebildet hat und diese eine Firma gründen, muss man ihnen etwas zutrauen und sie unterstützen. Deshalb habe ich sie gefragt, ob sie ein paar meiner Schriften mit Serifen verkaufen wollen. Sie haben ja gesagt. Mit der BigVesta[11] bin ich zu Linotype[12] gegangen. Einfach um zu sehen, ob es Unterschiede gibt. Aber es dauert Jahre, bis man das vergleichen kann. Der Hintergrund ist also, ich will nicht alles alleine machen, ich will ein Teamplayer sein.

Sie haben einige Namen erwähnt. Veronika Burian ist in der Ausstellung »Call for Type« vertreten, Timo Gaessner[13] war Ihr Schüler. Wie sind Schriftgestalter untereinander vernetzt?

Wir können im Allgemeinen sehr gut miteinander reden. Ich hatte bisher kaum Probleme mit Kollegen. Wir treffen uns öfter auf Konferenzen und sind alle gute Freunde. Ab und zu betonen Kollegen: »Das ist von mir und da muss man die Finger davon lassen.« Dann gibt es komische Diskussionen. Unabhängig davon ist es manchmal einfach so: Aus verschiedenen Anregungen an unterschiedlichen Orten und Zeiten können für bestimmte Probleme vergleichbare oder dieselben Lösungen entstehen. Die Probleme sind oft gleich und es liegt bei der Schriftgestaltung schon so viel fest, dass Schriftgestalter auf dieselben oder vergleichbare Lösungen stoßen.

Sie unterrichten, leiten Workshops, vermitteln Ihr Wissen weiter. Was raten Sie, wenn es darum geht, ein Gefühl für Typografie zu entwickeln?

Wahnsinnig viel lesen. Und nicht nur den Text, sondern auch die Buchstabenformen anschauen. Denn speziell, wenn man denkt, dass sich etwas sehr angenehm liest, sollte man direkt nachforschen, woran das liegt. Und weiter: Man sollte sich sehr breit orientieren. Die Anregungen für neue Schriften können überall herkommen.

Haben Sie im Gegenzug auch etwas von Ihren Studierenden und Schülern gelernt?

Unterrichtend tätig sein, ist die beste Möglichkeit selbst viel zu lernen. Ich lerne genauso viel wie meine Studenten. Wenn ich meinen Studenten eine Frage stelle und sie auffordere, für ein bestimmtes Problem eine Lösung zu finden, dann kommen ab und zu Lösungen heraus, von denen ich denke, ich hätte sie eigentlich selbst finden müssen.

Außerdem denken Studenten anders über Schrift nach, weil sie viel jünger sind und die Welt anders sehen, weil sie andere Erfahrungen haben als ich. Nicht so viel historischen Ballast. Sie gehen einfach anders mit Schrift um. Da realisiere ich, dass ich das früher auch so gemacht habe. Erfahrung ist fantastisch, aber man muss immer aufpassen, dass man sich nicht wiederholt oder Lösungen benutzt, die man schon früher verwendet hat. Ehrlich gesagt, bin ich nur froh, dass viele Menschen es völlig anders machen als ich, dann gibt es gute Diskussionen und Fragen, die ich mir stelle; ich bleibe wach und interessiert. Ansonsten wäre es wahnsinnig langweilig in der Welt und glücklicherweise ist es das nicht.

Das Interview mit Gerard Unger führten Tabea Dölker und Julia Bielefeld.

9 Veronika Burian (→ S. 32)

10 José Scaglione (*1974) ist ein argentinischer Grafik- und Type-Designer, der an den Universitäten in Rosario und Buenos Aires in Argentinen unterrichtet. Darüber hinaus ist er seit 2007 Mitglied des Vorstandes des »Association Typographique Internationale«. Er gründete 2006 zusammen mit Veronika Burian die Type Foundry Type Together (→ S. 32, 126).

11 BigVesta (2003) ist eine Display-Schrift von Gerard Unger, die in kleinen Größen wie auch als Textschrift funktioniert.

12 Die 1890 gegründete Monotype GmbH vertreibt bzw. lizensiert heute hochwertige digitale Schriften über monotype.com Außerdem entwickelt Monotype Corporate Fonts für Geschäftskunden und stellte 2005 das Font Management-Programm Font Explorer zur Verfügung (→ S. 51, 80).

13 Timo Gaessner (→ S. 150)

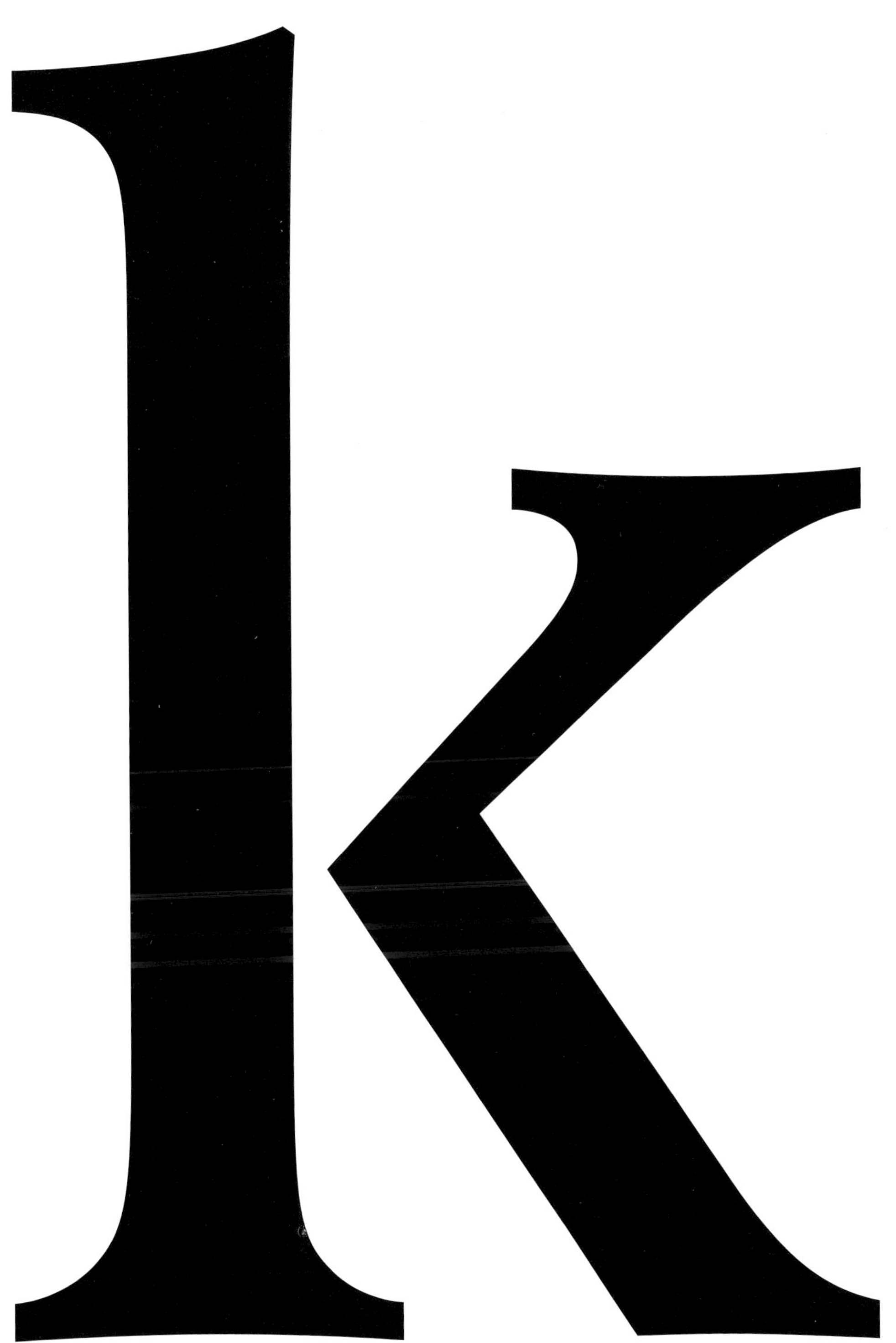

GERARD UNGER

Gerard Unger[1] (*1942) studied at the Gerrit Rietveld Academie until 1967 and started work as a freelance designer and type designer in 1972. He has taught at the Rietveld Academie for thirty years and currently holds professorships in Reading,[2] England and Leiden, Netherlands. His publications received several international awards, e.g. the H. N. Werkman Prize and the Piet Zwart Prize 2012. Unger's publication »Terwijl je leest«[3] (»While you're reading«), which deals with the reading process, has been translated into many languages including Korean.

How did you get interested in type design?

gu: I wasn't a particularly good student until I came to the Gerrit Rietveld Academie. I've always had problems reading. A colleague of mine considered my career choice as a compensation strategy. During primary and secondary school things didn't quite pan out. At Rietveld, I suddenly discovered that I was actually able to get something going. When I realized that I could personally add something to the world, I enjoyed this moment so much – and I enjoy it even now – that I sit in front of my computer every day.

What does your process of designing type look like? What materials do you use? Where do you begin?

Since 1986 when I bought my first Mac, I haven't drawn by hand anymore. Before that, I worked by means of a light table and tracing paper and drew fine pencil lines. When I began to use a Mac, I quickly noticed that working became much more direct. You have a computer and a printer; you create forms and you can make a direct print. Then you have a look at it, notice that something has to be changed, and continue your work. Every now and then, I still make some sketches. This happens during a kind of break when I don't get very far with my work and can't see any more possibilities – in my head, not on the computer. Actually the computer makes everything possible. For example, I can change each of my curves or leave out the serifs. All this can be done very quickly, and you can spend a lot of free time doing such things.

Your fonts are used for newspapers, magazines, road signs, and books. Do you know right away what purposes they will serve?

I could plan on creating a font for a newspaper. But then some users would see it and say: »That's a wonderful typeface for a book«. Others would say: »That's an excellent corporate identity font!« If one of my students said: »I would like to create a special font for a children's encyclopaedia«, then I would say: »Okay, go ahead. But you should be aware that the users of your font will use it for totally different purposes.« Because there is gap between designer and user. Actually I always keep the idea in mind that a font shouldn't be intended for any special purpose but it should have a wide range of applications. That's just the way it is.

Sometimes you also use a paintbrush and conduct some experiments. How much influence does coincidence have on your fonts?

There's always some kind of coincidence. There are moments in a creative process when all of a sudden everything goes well and develops automatically. When your work is done, you're astonished and wonder whether you really did it on your own. These moments actually contain a lot of experience; experience plays a certain role, too. And you just get carried away by ideas, experience, and visions. This also happened to me when I developed Swift:[4] I knew right away what I wanted to create. But when I made my sketches, I suddenly noticed that something was going to happen, and so I quickly continued my work. I went on with it for almost 24 hours at a stretch and this without any sleep or food. However, with regard to several other fonts, I actually had to force myself to develop them. Sometimes you have to put your results into a drawer for a month, and only then can you go on with your work.

How important is it for you to cooperate with others in the field of type design?

Type design is, if you like, a very lonely occupation. I always conferred with some experts in this field, such as newspaper designers, even when I created fonts on my own behalf like Swift, for example. I wanted to know their wishes and what they thought about several things. As a type designer, I've always tried to socialize as much as possible. It means loneliness when you sit at the computer for hours, and I can't stand it for long.

I cooperated with many former students in order to create my last font, which has an elaborate character set of almost several thousand signs. Among other things, it contains polyphonic Greek, and possibly Korean, Japanese, Arabic, and Devanagari will be added to it some time or other. I also work with former students like Tom Grace from Heidelberg or Irene Vlachou, both former students at Reading, in the field of type production. I create the basic set of signs, and the two of them work out them. He develops small caps, text figures etc., and combinations of signs for many languages; she assists me with Greek. When I get the elaborated signs back, I can say: »No, this curve will need some rework so that it looks again like the curve I had in mind.«

How did you come up with the names of your fonts such as Gulliver[5] or Swift?

This is the most difficult thing. Matthew Carter,[6] a colleague of mine, said: »I would rather create ten new fonts than come up with the name for one of them«.

1 gerardunger.com

2 Reading University is in Reading, UK. It was founded in 1926 and is famous for its Master Programme for Type design. Many of the designers mentioned in this publication have studied or taught there (→ S. 184).

3 Gerard Unger's »While You're Reading« deals with the reading process. The book has been translated into many languages. Unger, Gerard: While you're reading. New York: Mark Batty, 2007.

4 Gerard Unger developed the Swift between 1984 and 1987 as a modern digital typeface for newsprint.

5 The Gulliver is a space-saving and very legible typeface for newsprint, created by Gerard Unger in 1993. The exclusive licence is limited to 100 only.

6 Matthew Carter is an English type designer with more than 50 years' experience, during which time he worked for several Linotype companies in the US and in Europe. In 1981 he was co-founder of the type foundry Bitstream Inc., Cambridge, Mass., which he left in 1991 together with Cherie Cone in order to establish Carter & Cone Type, Inc.

For several fonts I had suitable names in mind even before they were ready. This happened when I created Swift. During my work on this, there were always swifts hovering above me in the sky and cheering me on. They served as an inspiration. With regard to Gulliver, it took a bit more time to find a name for this font. When I experimented with it, I realized that I was about to develop a font which looked much bigger in its very small sizes than it actually was and that you could use it in much smaller sizes than it was possible for many other fonts. So I knew that this font had to be called »Gulliver«. Gulliver was a giant among Lilliputians and a dwarf among giants. What a perfectly suitable name! However, it took me a lot of time for several other fonts.

You sell Capitolium 2[7] through TypeTogether.[8] How did this collaboration come about?

I have sold my fonts for more than ten years. When the orders arrived in my studio, I dealt with them on the same day. At the beginning, I posted floppy discs, later I simply sent e-mails. But I had the same old problem with loneliness. If you are occupied with orders all day long and at the same time you try to do your work as type designer, graphic designer, and typographer, you sooner or later have to delegate some of your work. Veronika Burian[9] and José Scaglione[10] were both former students at Reading. I thought: you trained these students, and they set up an enterprise. Now you have to believe in them and support them. For that reason I asked them whether they wanted to sell some of my serif typefaces, and they accepted my offer. With regard to BigVesta,[11] I involved Linotype.[12] I just wanted to find out whether there was any difference. But it takes a long time before you can compare the outcome. Well, this is what actually led to the idea of cooperation: I don't want to do everything on my own; I want to be a team player.

You have mentioned the names of several persons. Veronika Burian is represented at the »Call for Type« exhibition. Timo Gaessner[13] was one of your students. What does the network among type designers look like?

In general, we get along very well. I've rarely had any problems with my colleagues so far. We occasionally meet at symposia and are all good friends. Sometimes one of my colleagues says intensely: »It was me who made that and you must keep your hands off it!« This can lead to rather peculiar discussions. On the other hand, things can sometimes just develop in the same way. Certain problems can lead to the same or similar solutions because of various stimuli at different places and different times. Actually in the field of type design the problems are often the same, and many things have already been determined so that type designers can easily come across the same or similar solutions.

You are an instructor, you run workshops, you share your knowledge. In your opinion, what should people do to get a feel for typography?

A lot of reading, and you shouldn't only look at the text but also at the letter forms. Especially if you think that a text is really easy on the eyes, you should immediately try to find out the reason for it. And more than that: you should be open to many fields of interest. You can get an idea for the creation of a new font everywhere you go.

Did you also learn anything from your students and course participants?

Teaching offers the best chance of learning a lot. I learn as much as my students do. When I put a question to my students and ask them to find a solution for a certain problem, they occasionally come up with ideas which make me think that actually I should have found such solutions on my own.

In addition to that, students reflect on typefaces in a different way because they are much younger. Since they have gained a different kind of experience, they look at the world differently. There's not so much historical burden. They just take a different approach to type design. However, I realize that I used to do the same. Experience is fantastic but you must always make sure you don't repeat yourself or go for solutions you have used before. Well, to be perfectly honest, I'm rather glad that many people use working methods that are completely different from mine. This leads to good discussions and makes me think about various questions. All this keeps me alive and interested. Otherwise the world would be an awfully boring place, and fortunately that's not the case.

Gerard Unger was interviewed by Tabea Dölker and Julia Bielefeld.

7 The Capitolium was created in 1998 for the jubilee of the Roman-catholic church as a typeface for a guiding and information system in Rome. A revised version for newsprint came out in 2006 as Capitolium 2 or Capitolium News, resp.

8 José Scaglione and Veronika Burian met during their Master Programme at Reading University, and in 2006 they established the foundry »TypeTogether« (→ S. 37).

9 Veronika Burian (→ S. 37)

10 José Scaglione (*1974) is an Argentinian graphic and type designer teaching at the Universities of Rosario and Buenos Aires. Since 2007 he has also been a Board Member of the Association Typographique Internationale. In 2006 he founded, together with Veronika Burian, the type foundry TypeTogether (→ S. 37, 129).

11 Big Vesta (2003) is a display type face by Gerard Unger, which works in small sizes as well as for text.

12 Monotype GmbH, established in 1890, sells and licences high quality digital fonts via monotype.com. Monotype also develops corporate fonts for clients and in 2005 made the Font Management Programme Font Explorer available (→ S. 54, 85).

13 Timo Gaessner (→ S. 155)

KARL NAWROT

PAUK, MONA, JOSA & BREU.

ABCDEFGHIJKLMN
OPQRSTUVWXYZ

Dess Pauk

ABCDEFGHIJKLMN
OPQRSTUVWXYZ

Dess Mona

ABCDEFGHIJKLMN
OPQRSTUVWXYZ

Dess Josa

ABCDEFGHIJKLMN
OPQRSTUVWXYZ

Dess Breu

Schrift. Typeface. Dess
Gestalter. Designer. Karl Nawrot
Label. Foundry. voidwreck.com
Jahr. Year. 2012

Die Dess-Schriftfamilie wurde 2012 vom Grafik-Design-Studio Our Polite Society (Amsterdam, NL) für die neue Identität der Stiftung Bauhaus Dessau in Auftrag gegeben. Die Familie besteht aus vier Schriften (Josa, Breu, Mona und Pauk) und wurde von vier verschiedenen Persönlichkeiten des Bauhauses inspiriert: Josef Albers, Marcel Breuer, László Moholy-Nagy und Paul Klee.

The Dess Family was commissioned by the graphic design studio Our Polite Society (Amsterdam, NL) and used for the new identity of the Bauhaus Dessau Foundation in 2012. The family consists of four typefaces (Josa, Breu, Mona and Pauk), and was inspired by four different personalities of the Bauhaus – Josef Albers, Marcel Breuer, László Moholy-Nagy and Paul Klee.

Der französische Grafik-Designer *Karl Nawrot*[1] (*1976), dessen Arbeit zwischen Illustration, Schriftgestaltung und Plastik oszilliert, betreibt mit seinem Alter ego Walter Warton die Website »Voidwreck«. Er studierte Illustration in Lyon (FR) und absolvierte 2008 seinen Master an der Werkplaats Typografie in Arnheim. 2008 bis 2012 lehrte Nawrot Zeichnen an der Gerrit Rietveld Academie, zurzeit unterrichtet er an der University of Seoul.

Sie studierten in Lyon Illustration, ehe Sie an einem Programm beim Werkplaats Typografie in Arnheim teilnahmen, und wenn man sich Ihr Portfolio ansieht, dann merkt man, dass Sie mit einer ganzen Reihe von Ausdrucksformen vertraut sind. Vor allem Schriftdesign (sowohl gezeichnet als auch digital), Modellbau (Gebäude und typografische Module) und Illustrationen (architektonische Darstellungen und abstrakte Zeichnungen). Wie würden Sie Ihren Ansatz heute beschreiben?

kn: [...]

In der Ausstellung »Call for Type« wird Ihre Schriftfamilie Dess gezeigt, die vom Graphic Design Studio »Our Polite Society«[2] kommissioniert wurde und 2012 für das Erscheinungsbild der Stiftung Bauhaus Dessau[3] benutzt wurde. Die vier verschiedenen Schriften basieren auf vier verschiedenen Persönlichkeiten des Bauhauses – Josef Albers,[4] Marcel Breuer,[5] László Moholy-Nagy[6] und Paul Klee.[7] Warum haben Sie gerade diese vier für Ihre Schriftfamilie Dess gewählt?

Die ersten drei wurden von »Our Polite Society« ausgewählt. Nur für die letzte Schrift durfte ich zwischen Paul Klee und Wassily Kandinsky[8] wählen. Ich entschied mich für den, der mir näher stand.

Welcher Aspekt der Arbeiten von Albers, Breuer, Moholy-Nagy und Paul Klee beeinflusste den Arbeitsprozess der Dess-Familie und ihre Schriften Josa, Breu, Mona und Pauk?

Josa basiert auf einer Sammlung von Schablonenfonts von Josef Albers. Ich wollte aber keinen weiteren »Hommage-Font« schaffen, also schnitt ich mir selbst Schablonen, um die Buchstaben zu zeichnen. Die digitale Version der Schrift besteht aus einer Auswahl der Glyphen, die gezeichnet oder aus der Praxis abgeleitet sind.

Breu wurde von Marcel Breuers brutalistischer Architektur inspiriert. Die Hauptidee für den Font bekam ich von einem verlassenen Gebäude, das Breuer konzipiert hatte – dem Parador Ariston. Als ich mir die Fotos mit den Innenansichten ansah, hatte ich den Eindruck, dass ich keine Räume sah, sondern eher Höhlen. Ehe ich anfing, den Font zu zeichnen, machte ich mir zwei Gipsmodelle von Breuers

1 voidwreck.com

2 Our Polite Society ist ein Studio in Amsterdam und Stockholm, das 2008 von Jens Schildt (SE) and Matthias Kreutzer (D) gegründet wurde. Arbeitsschwerpunkt bilden Aufträge im Bereich Printmedien, aber auch die Gestaltung von Websites ist Bestandteil ihrer typografischen Arbeiten.

3 Die Stiftung Bauhaus Dessau beschäftigt sich mit der Erhaltung, Erforschung und Vermittlung des Bauhauserbes und versteht sich als ein Ort experimentellen Gestaltens. Ergebnisse der Arbeit werden u.a. durch Ausstellungen, Publikationen und Konferenzen kommuniziert.

4 Josef Albers (1888–1979), deutscher Maler, Bildhauer und wichtiger Bauhauslehrer. Nach der Schließung des Bauhauses durch die Nationalsozialisten musste er in die USA emigrieren und lehrte am Black Mountain College in North Carolina. Im typografischen Bereich ist Albers besonders für seine auf geometrischen Grundformen basierende Kombinationsschrift bekannt (→ S. 168).

5 Marcel Breuer (1902–1988) ungarisch-deutsch-amerikanischer Architekt und Designer, lernte und lehrte am Bauhaus. Er wurde v.a. durch seine Stahlrohr-Möbel wie dem Stuhl B3, auch Wassily-Chair genannt, bekannt.

6 László Moholy-Nagy (1895–1947), ungarischer Maler, Typograf und Fotograf. Bekannt ist er u.a. für seine konstruktivistische Malerei und sein typografisches Schaffen, seine Beschäftigung mit dem experimentellen Film am Staatlichen Bauhaus. Er sah die Fotografie als ideales Ausdrucksmittel, mit dem er die Theorie des »Neuen Sehens« umsetzen wollte.

7 Paul Klee (1879–1940), Schweizer Maler, Grafiker und Bauhausmeister.

8 Wassily Kandinsky (1866–1944), russischer Maler und Grafiker sowie Bauhausmeister.

Konstruktionen, die zeigten, dass seine Architektur gleichzeitig ein Nest und eine Höhle darstellte. Und diesen Eindruck illustriert der Font.

Mona fing mit dem Bild des Licht-Raum-Modulators[9] an, den László Moholy-Nagy entwarf. Die ersten Zeichnungen des Fonts bestanden aus einer Reihe von Collagen aus verschiedenen Komponenten, die von dieser Maschine inspiriert waren, sie wurden auf Glasplatten gesetzt und auf Polaroid-Bildern festgehalten. Mit diesen Experimenten war ich dann imstande, die ersten Entwürfe der Schrift zu zeichnen. Schließlich löschte ich die Punkte und schuf aus zwei vorherigen Arbeiten einen neuen Satz von Elementen, mit denen mir dann der endgültige Font gelang.

Pauk basiert auf Paul Klees berühmtem Zitat »Eine Linie ist ein Punkt, der spazieren geht«. Jeder Buchstabe wurde zu einem Plan der Route: die Blöcke waren Hindernisse und die gepunktete Linie natürlich der Weg.

Wenn man von Ihrem Hintergrund ausgeht, der sowohl Schriftdesign als auch Illustration beinhaltet, glauben Sie, dass diese beiden Disziplinen einander beeinflussen?

Ja, natürlich.

Sie haben von 2008 bis 2012 auch an der berühmten Gerrit Rietveld Academie unterrichtet. Haben diese vier Jahre als Lehrer Ihre Einstellung zu Schriftdesign beeinflusst?

Nein, ich gab ja Zeichenunterricht.

Nach Ihrer Lehrtätigkeit an der Rietveld Academie gingen Sie nach Südkorea, wo Sie an der Universität von Seoul unterrichteten. Zwei Schulen in zwei sehr verschiedenen Ländern, jedes mit einer völlig anderen Kulturgeschichte. Wie stark nimmt man diese Unterschiede im Schriftdesign wahr?

Der einzige Unterschied liegt in den Schriften: hier Lateinisch, dort Hangul.

Mit der Erfindung des PC bekamen wir auch das digitale Schriftdesign. Im Gegensatz hierzu scheint Ihre Arbeit eher analog und handschriftlich zu entstehen. Haben Sie den Eindruck, dass Sie von den Fortschritten der digitalen Technik profitiert haben, oder halten Sie diese für Ihren Ansatz im Schriftdesign für überflüssig?

Die Frage ob analog oder digital interessiert mich nicht – ich bemühe mich lediglich, für jedes Projekt, an dem ich gerade arbeite, um den bestmöglichen Weg.

Hat Ihre augenblickliche Arbeit etwas mit Typografie zu tun?

Nein, ich beschäftige mich mit ganz anderen Dingen.

Das Interview mit Karl Nawrot führten Robin Scholz und Felix Rank.

9 Der Licht-Raum-Modulator ist eine kinetische Plastik, die László Moholy-Nagy als »Apparat zur Demonstration von Licht- und Bewegungserscheinungen« bezeichnet hat.

Dess Breu

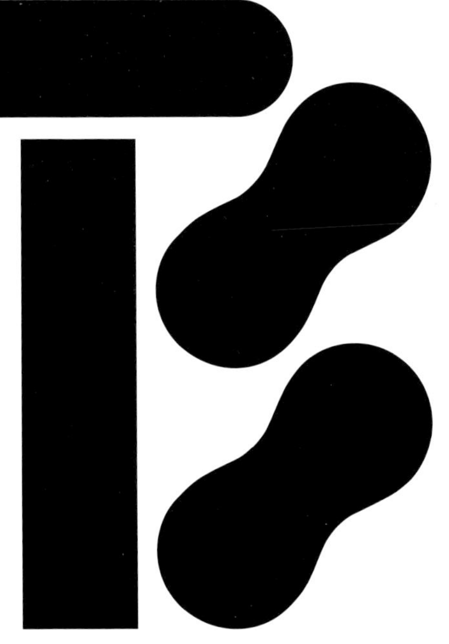

Dess Josa

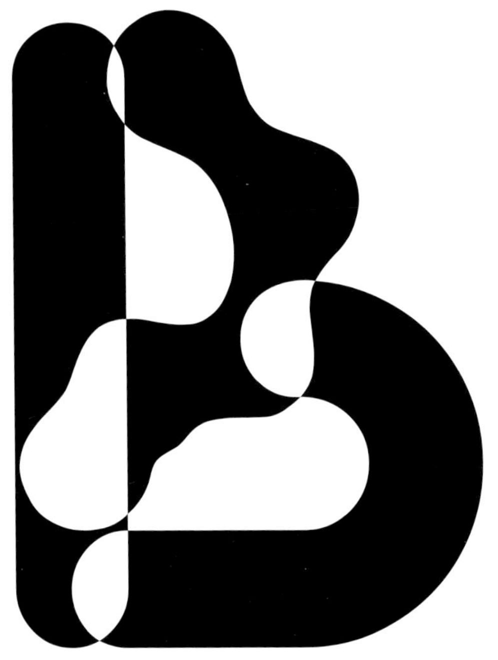

Dess Mona

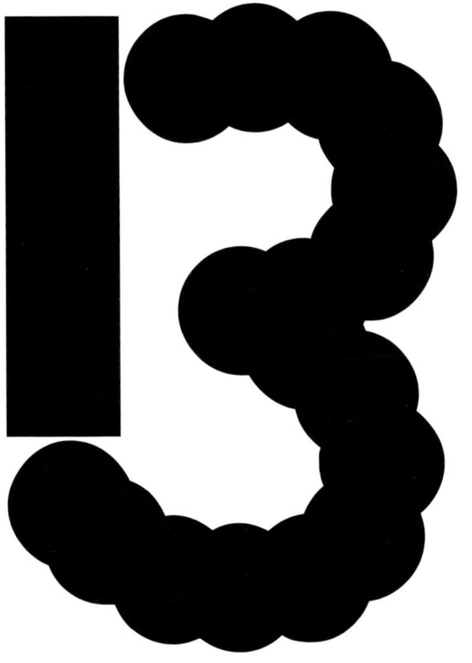

Dess Pauk

***Karl Nawrot*[1] (*1976), a French graphic designer, whose work oscillates between illustration, type design and sculpture, runs the website »Voidwreck« together with Walter Warton, his alter ego. He studied Illustration in Lyon, France, and in 2008 obtained his Master Degree at Werkplaats Typografie at Arnhem. Nawrot taught drawing at the Gerrit Rietveld Academie from 2008 until 2012 and is currently teaching at the University of Seoul.**

You studied illustration in Lyon before participating in a MA residential programme at the Werkplaats Typografie in Arnhem, and by taking a look at your portfolio it seems that you embrace quite a few forms of expression, mostly type design (both hand-drawn and digital), model making (buildings and typographic modules) and illustrations (architectural representations and abstract drawings). How would you describe your own approach today?

kn: […]

The »Call for Type« exhibition will present the Dess Family which was commissioned by the graphic design studio »Our Polite Society«[2] and used for the 2012 identity of the »Bauhaus Dessau Foundation«.[3] The four different typefaces are based on four different personalities of the famous Bauhaus – Josef Albers,[4] Marcel Breuer,[5] László Moholy-Nagy[6] and Paul Klee.[7] Why did you choose these four as the foundation for the Dess Family?

The first three were selected by »Our Polite Society«. Only for the last one, I had to choose between Paul Klee and Wassily Kandinsky.[8] I just picked the one I felt closer to.

Which particular portion of the work of Albers, Breuer, Moholy-Nagy and Paul Klee influenced the design process of the Dess Family and its typefaces Josa, Breu, Mona and Pauk?

Josa was based on the collection of Josef Albers' stencil fonts. As I didn't want to make another homage font, I cut my own templates to draw the letters. The digital version of the typeface is a selection of the glyphs that had been drawn with the tool or were deduced from its practice.

Breu was inspired by Marcel Breuer's brutalist architecture. The main idea behind the font came from an abandoned building designed by Breuer – The Parador Ariston. After looking at the photographs of the inside spaces I felt that I was not looking at rooms anymore but simply caves. Before starting to draw the font I made two plaster models of Breuer's constructions that showed his architecture playing the roles of a nest and a cave simultaneously. In the end the font just illustrated that idea.

Mona started with the picture of the Light Space Modulator[9] designed by László Moholy-Nagy. The first sketches of the font were a series of collages built with different components inspired by his machine, mounted on slides and then captured on Polaroids. These experiments allowed me to draw the first draft of the typeface. I finally removed the dots and created a new set of elements taken from two previous jobs and I used them to achieve the final font.

Pauk was based on Paul Klee's famous quote – A line is a dot that went for a walk. Each letter was drawn as a map of a route: the blocks were obstacles and the line of dots, obviously, the walk.

Regarding your background in type design as well as in illustration, do you feel that these two fields within your work influence each other?

Yes, obviously.

You have also taught at the famous Gerrit Rietveld Academie from 2008 until 2012. Did the four years as a teacher influence your approach to type design?

No, I was teaching drawing.

After teaching at the Rietveld Academie you moved to South Korea to teach at the University of Seoul. Two schools representing two different countries which vary quite a bit in their cultural history. How visible are these differences in the respective approach to type design?

The only difference is the alphabets: Roman versus Hangul.

With the invention of the personal computer, digital type design followed suit. In contrast, your work seems rather tool and analog based. Do you feel that you benefitted from the digital advancements at all, or do they seem rather unnecessary to your approach in type design?

I have no interest in the analog or digital question – I just want to find the right process for each project I happen to be working on.

Anything type-related you are currently working on?

No, I am following some other interests.

Karl Nawrot was interviewed by Robin Scholz and Felix Rank

DESS

1 voidwreck.com

2 Our Polite Society is a studio in Amsterdam and Stockholm which was founded in 2008 by Jens Schildt (SE) and Matthias Kreutzer (D). Emphasis is on work in the area of print media, but website design is also part of their typographic work.

3 The Bauhaus Dessau Foundation is concerned with the upkeep, research and the conveying of the Bauhaus heritage and sees itself as a place for experimental design. Results of the work are shown in exhibitions, in publications and at conferences.

4 Josef Albers (1888–1979), German painter, sculptor and important Bauhaus teacher. After the Bauhaus was closed by the Nazis he emigrated to USA, where he taught at the Black Mountain College in North Carolina. In typography, Albers is well-known for his combination type, based upon geometric principles (→ S. 173).

5 Marcel Breuer (1902–1988), Hungarian-German-American architect and designer, trained and taught at the Bauhaus. He became well-known for his steel tube furniture, e.g. the chair B3, also known as Wassily chair.

6 László Moholy-Nagy (1895–1947), Hungarian painter, typographer and photographer. He became famous for his constructivist painting and his work in typography, also for his work in experimental film at the Staatliches Bauhaus. He saw photography as an ideal expressive medium, with which he wanted to implement the theory of »A new Way of Looking«.

7 Paul Klee (1879–1940), Swiss painter, graphic artist and teacher at the Bauhaus.

8 Wassily Kandinsky (1866–1944), Russian painter and graphic artist, teacher at the Bauhaus.

9 The Light Space Modulator is a kinetic sculpture which László Moholy-Nagy called »An apparatus for the demonstration of phenomena of light and movement«.

LARS HARMSEN & BORIS KAHL

Schrift. Typeface.	Fraktendon Pro
Gestalter. Designer.	Lars Harmsen & Boris Kahl
Label. Foundry.	volcano-type.de
Jahr. Year.	2003

Fraktendon entstand 2003 aus der einfachen Idee, Stilmerkmale zweier völlig unterschiedlicher Schriften zu verbinden. Aus Fraktur und Clarendon wurde Fraktendon. Weit hergeholt ist die Kombination dieser beiden Schriften nur auf den ersten Blick: Die Clarendon erfreute sich im ausgehenden 19. Jahrhundert auch im deutschsprachigen Raum großer Beliebtheit und war oft Seite an Seite mit gebrochenen Schriften zu sehen. Damals aber »wohlgesittet« nebeneinander – nicht miteinander gekreuzt. Das, was mit der Fraktendon ursprünglich als plakatives Statement – Neues entsteht durch Mischung – konzipiert war, wurde unverhofft zu einem Bestseller. Zum 10-jährigen Jubiläum der Fraktendon erscheint Ende 2013 die Fraktendon Pro, eine überarbeitete Version, die auf den aktuellen Stand der Technik gebracht und mit zahlreichen OpenType-Funktionen und Sonderbuchstaben ausgestattet wird.

Fraktendon was developed in 2003 from the simple idea of combining the style characteristics of two completely different typefaces. Fraktur (black letter type) and Clarendon were combined and became Fraktendon. This seems far-fetched only at first sight: Clarendon enjoyed great popularity not only in Germany at the end of the 19th century and often was seen side by side with broken script. In those days, however, well behaved and side by side – not mixed up.

But the idea originally conceived as a bold statement – new things happen by mixing the established – unexpectedly turned into a bestseller. And to mark the 10th anniversary of the Fraktendon, we shall have, by the end of 2013, the Fraktendon Pro, a revised version, technically updated and including numerous OpenType functions and additional characters.

DNA
Heterosis-Effekt
Bastard
Hybrid-Zucht
Inzucht
Mischling

FRAKTENDON PRO

Der als Professor für Konzeption, Entwurf / Typografie und Layout an der FH Dortmund, Fachbereich Design, lehrende *Lars Harmsen*[1] (*1964) ist Geschäftsführer der Agentur »MAGMA Brand Design«[2] (Karlsruhe) sowie Gesellschafter und Creative Director von »Melville Brand Design«[3] (München). 1996 gründete er mit Uli Weiß[4] den Schriftenverlag »Volcano Type«,[5] der derzeit über 200 Schriften vertreibt. 2004 rief Harmsen den Blog »Slanted«[6] ins Leben, 2005 das gleichnamige Typo-Magazin. Prof. Harmsen ist Mitglied im ADC e.V. und Autor bzw. Gestalter zahlreicher Publikationen, darunter »BASTARD – Choose my Identity«, »TYPODARIUM« und »Yearbook of Type«.

Woher kommt Dein Interesse für Typografie? Seit wann beschäftigst Du Dich mit Schrift, schon im Studium oder erst danach?

lh: Mein Interesse hat bedingt im Studium begonnen. So viel typografische Ausbildung gab es leider nicht, als ich studierte (Studienabschluss 1996). Ich hatte ganz am Ende des Studiums einen Professor, der die ersten Emigre-Magazine von Rudy VanderLans[7] in den Unterricht mitgebracht hat. Das war für uns, wie man so schön sagt, »Mindblowing«. Das hat das Fieber ausgelöst! Schriftenmachen wurde möglich. Damit gestalten war aufregend. Die ersten Fonts von Emigre waren Pixel-Schriften wie die Oakland,[8] also nichts, was man nicht auch selber hätte machen können. Während des Studiums habe ich zudem viel für die Plattenindustrie und Modebranche gearbeitet, u.a. für Fanatic, ein Surfboard- und Sportswear-Brand. Von Emigre und Surfen war der Weg nicht weit zum Ray Gun-Magazin von David Carson.[9] So sind dann auch die ersten Fonts entstanden: einfach losgelegt. Die Platten-Cover waren in erster Linie für Techno-Produktionen. Das Bild/Foto hat dabei keine so große Rolle gespielt, es ging eher um Typografie. Zudem gab es Photoshop zu diesem Zeitpunkt noch nicht. Alles, was man am Computer (SE 30) machen konnte, war Schriften hin- und herschieben. Diese Zeit hat mich sehr geprägt. Man hat mit wenig Mitteln viel versucht.

1 magmabranddesign.de

2 1996 gründeten Lars Harmsen und Uli Weiß die Karlsruher Design-Agentur Magma Brand Design.

3 Das 2008 von Michael Schmidt, Florian Brugger, Johannes König, Lars Harmsen und Ulrich Weiß gegründete Münchner Grafik-Design-Büro Melville Brand Design ist die Partneragentur von Magma Brand Design.

4 Geschäftsführer von Magma Brand Design.

5 1996 gründete Magma Brand Design die deutsche Type Foundry Volcano Type, die neben experimentellen Display-Fonts heute auch große Schriftfamilen wie die Matryoshka oder die Telegramo verlegt.

6 Der 2004 gegründete Weblog slanted.de widmet sich typografischen und gestalterischen Entwicklungen und Trends. Inzwischen ist slanted.de zu einem der wichtigsten Design-Portale und Diskussionsforen im deutschsprachigen Raum herangewachsen. Das 2005 gegründete Magazin SLANTED ist monothematisch ausgerichtet und widmet sich speziellen typografischen Themen.

7 Rudy VanderLans (*1955 Voorburg) ist ein niederländischer Schrift- und Grafik-Designer, der ab 1984 zusammen mit seiner Frau Zuzana Licko das Emigre Magazine → S. 43, eine Zeitschrift für experimentelles Grafik-Design, veröffentlichte.

8 Oakland ist eine Reihe aus vier Pixel-Schriften mit unterschiedlicher Pixelhöhe für niedrig auflösende Display-Darstellung. Die von Zuzana Licko im Jahr 1985 entworfene Schrift wurde bei Emigre veröffentlicht.

9 Der US-amerikanischer Typograf, Grafiker und Surfer David Carson (*1957) wurde bekannt durch seine innovative Magazingestaltung und die Verwendung experimenteller Typografie. Sein Typografie- und Layout-Stil prägte die Grunge-Ästhetik. Carson war Art Director der Zeitschrift Ray Gun, bevor er 1995 sein eigenes Studio in New York gründete (→ S. 179).

Du hattest schon vorab erwähnt, Du seiest unter den in der Ausstellung aufgeführten Personen am wenigsten der Type Designer – wie würdest Du Dich bezeichnen?

Die Schriften, die ich bisher gemacht habe, sind aus einer ganz anderen Motivation heraus entstanden, wie sicherlich viele Fonts, die in der Ausstellung »Call for Type« gezeigt werden. Meine Schriften mussten immer sofort zum Einsatz kommen. Der Auftrag für ein Cover, ein Magazin, ein T-Shirt war da, die Schrift dazu habe ich nicht gefunden, weil es diese einfach noch nicht gab. Also habe ich sie gestaltet. Ich kann mit Schriften gut umgehen, bin aber zu ungeduldig, um Schriftgestalter zu sein. Typografie entspricht mir da mehr.

Bist Du der Meinung, dass es typografische Trends gibt? Wie informierst Du Dich über typografische Entwicklungen?

Trends langweilen mich. Mit offenen Augen sieht man genug. Auch genug Mist und »Me-too«-Gestaltung. Aber auch sehr Gutes.

In der Ausstellung »Call for Type« hat man viele Fonts mit Ziehfeder-Ästhetik gesehen. In den Jahren zuvor waren es Grotesk-Bauhaus-Style-Konstruktionen. Davor charaktervolle Antiqua-Schriftentwicklungen. Ja, es gibt Trends. Aber muss man immer auf den Zug springen?

Wie sieht der Prozess des Schriftentwurfs bei Dir aus? Wie gehst Du bei der Schriftgestaltung vor?

Die Schriften, die ich bisher gemacht habe, waren oft konzeptionell. Der Einsatzbereich war durch den Auftrag definiert – eine Umsetzung mit sehr kurzen Entwicklungszyklen. Anfangs noch mit einer Software namens Ikarus. Ein Albtraum.

Was sind Deine Inspirationsquellen?

Reisen und Interdisziplinarität sowie Begegnungen mit anderen Menschen inspirieren mich. Daraus entstehen dann oft Ideen für ein Thema, gefolgt von einem Konzept. Wir haben z.B. eine ganze Schriftreihe entwickelt, die wir Bastard-Fonts genannt haben, weil wir ein Buch zum Thema Identität gemacht haben. Mit »Bastard – Choose my Identity« haben wir uns die Frage gestellt, wie entsteht Neues in einer Welt, in der schon so vieles definiert und ausprobiert wurde? Auf unserer Weltreise, zusammen mit einem Fotografen und einem Illustrator, haben wir festgestellt, dass im Kern schon vieles definiert worden ist, dass es relativ schwer ist, da wirklich Neues entstehen zu lassen, das Neue aber dadurch entstehen kann, dass man Dinge kombiniert und neu arrangiert. Wie das in der Musik schon lange der Fall ist. Die Grenzen sind nicht mehr so scharf zu trennen: Illustration, Typografie, Grafik-Design, Fotografie: Alles wurde vermischt – und daraus hat sich dann auch das Konzept der Fonts ergeben, die wir für das Buch entwickelt haben.

Gab es einen bestimmten Anlass für den Entwurf von Fraktendon Pro?

Die Fraktendon war die erste Schrift, die wir in der »Bastard-Fonts«-Reihe entwickelt haben. Den amerikanischen Charakter der Clarendon[10] haben wir mit einer deutschen Fraktur gekreuzt. Eine Versöhnung von weich und hart. Fraktendon wurde dann im Bastard-Buch und anderen Magazinen und Drucksachen eingesetzt.

Deine Vorbilder?

Ich mag Menschen, die unvoreingenommen Dinge einfach machen. Die nach vorne marschieren und etwas Neues ausprobieren. Nicht aus Respektlosigkeit vor dem, was da war, sondern aus Neugierde heraus.

10 Die Clarendon ist eine serifenbetonte Linear-Antiqua, die 1845 von Benjamin Fox entworfen und 1850 von der Bauerschen Schriftgießerei auf den Markt gebracht wurde. 1950 wurde sie von Hermann Eidenbenz über die Linotype wiederveröffentlicht. Auffällig sind die ausgeprägten Tropfenformen beispielsweise beim a, r oder g.

Jemand wie Lars von Trier[11] z.B., dessen Filme ich sehr inspirierend fand. Da gab es das Hollywood Kino – und er hat es ganz anders gemacht. David Carson hat auch Konventionen über Bord geworfen. Er hat ausprobiert, die Kritik der konservativen Füchse war ihm egal, er hat Tore für nachfolgende Generationen geöffnet. Ich mag Dinge, die nicht aus einer »verkopften« Überlegung heraus entstanden sind, sondern viel eher aus einem spontanen Impuls des Ausprobierens und des Entdeckens und die damit authentisch und greifbar sind.

Die meisten Projekte von Dir sind in Zusammenarbeit mit anderen entstanden, wie z.B. mit Boris Kahl[12] bei der Fraktendon Pro. Wie wichtig ist Dir Team-Arbeit?

Mit anderen zu arbeiten macht mir viel Spaß. Dialog und Austausch befruchten, lassen Irrwege erkennen. Gestalter aus anderen Disziplinen finde ich dabei besonders interessant. Wenn man mit Leuten arbeitet, die Dinge ganz anders wahrnehmen, entsteht oft eine positive Reibung. Das ist auch das Spannendste an »Slanted«. Immer im Dialog mit anderen Kreativen zu stehen, ist ein Katalysator für eigene Ideen. Für mich ist das eine Arbeitsmethode, die viel besser funktioniert, als im eigenen Kämmerchen zu hocken und auf den großen Wurf zu warten. Ich bin auch kein stiller Typ, reib' mich auch mal gerne, scheu' nicht die Diskussion um die Sache. Mal eingeschnappt sein, für ein paar Stunden den Mund halten, dann darüber nachdenken, um dann doch zu sagen: »Stimmt, das war ein ganz guter Einwand«, ist für mich okay. Mir tut es gut, mit anderen Menschen zusammenzuarbeiten!

Der Markt wird von Schriften überflutet – Free-Fonts unerwartet guter Qualität gibt es überall. Wie wird sich die Nachfrage nach professionell und in erster Linie von Gestaltern erstellten Schriften in Zukunft entwickeln?

Es findet seit Längerem eine Demokratisierung satt. Schriftenmachen bleibt nicht mehr einer kleinen Elite von Super-Spezialisten vorbehalten. Uns stehen genug Werkzeuge zur Verfügung, sowohl für die Gestaltung als auch für die Vermarktung. Heute kann letztendlich jeder, der sich ein wenig Mühe gibt, Schriften gestalten.

Im professionellen Bereich wird der Markt für Non-Latin-Fonts explodieren. Da wird es viel Bewegung und Entwicklungspotential geben, auch in finanzieller Hinsicht. Zudem werden Standards und Anforderungen immer höher. Als Gestalter will man auf gut ausgebaute Familien nicht verzichten müssen. Ligaturen sind Pflicht. Die Erwartungen der Gestalter sind in diesem Bereich viel größer geworden. Diesen Hunger können Free-Fonts nicht stillen.

Font-Labels, die es schaffen einen Gesamtcharakter zu erzeugen, wie ein gut funktionierendes Mode-Label – bei dem nicht nur der Schuh, sondern gleich die passende Trainingsjacke und die Mütze im Angebot sind – haben gute Marktchancen. Gerade für kleinere Schmieden ist das eine Chance. Man kauft dort gerne ein, denn oft passen die unterschiedlichen Schriften gut zusammen.

Welche Schriften kannst Du nicht mehr sehen?

Damals schon nicht, und heute auch nicht: Rotis.[13] Da bekomme ich Kieferschmerzen.

Das Interview mit Lars Harmsen führte Tobias Villmeter.

11 Lars von Trier (bürgerlich Lars Trier, *1956) ist ein dänischer Filmregisseur und Drehbuchautor. Er gilt als einer der markantesten und umstrittensten europäischen Filmemacher der Gegenwart. Neben seinen Spielfilmen (u.a. Antichrist) drehte Lars von Trier auch Werbespots und Musikvideos.

12 Boris Kahl (*1975) studierte Visuelle Kommunikation an der Hochschule Pforzheim. Seit 2001 ist er Art Director bei MAGMA Brand Design (Karlsruhe), realisierte zahlreiche eigene Schriftentwürfe und leitet den 1996 gegründeten Schriftenverlag Volcano Type (volcano-type.de).

13 Rotis ist eine Hybridschriftart, die 1988 von Otl Aicher → S. 152 veröffentlicht wurde. Die Schrift hat ihren Namen von Aichers Wohnort Rotis, einem Ortsteil von Leutkirch im Allgäu. Das Besondere der Schriftfamilie waren die bis dato unbekannten Varianten Semi-Antiqua und Semi-Grotesk sowie eigenwillige Einzelformen.

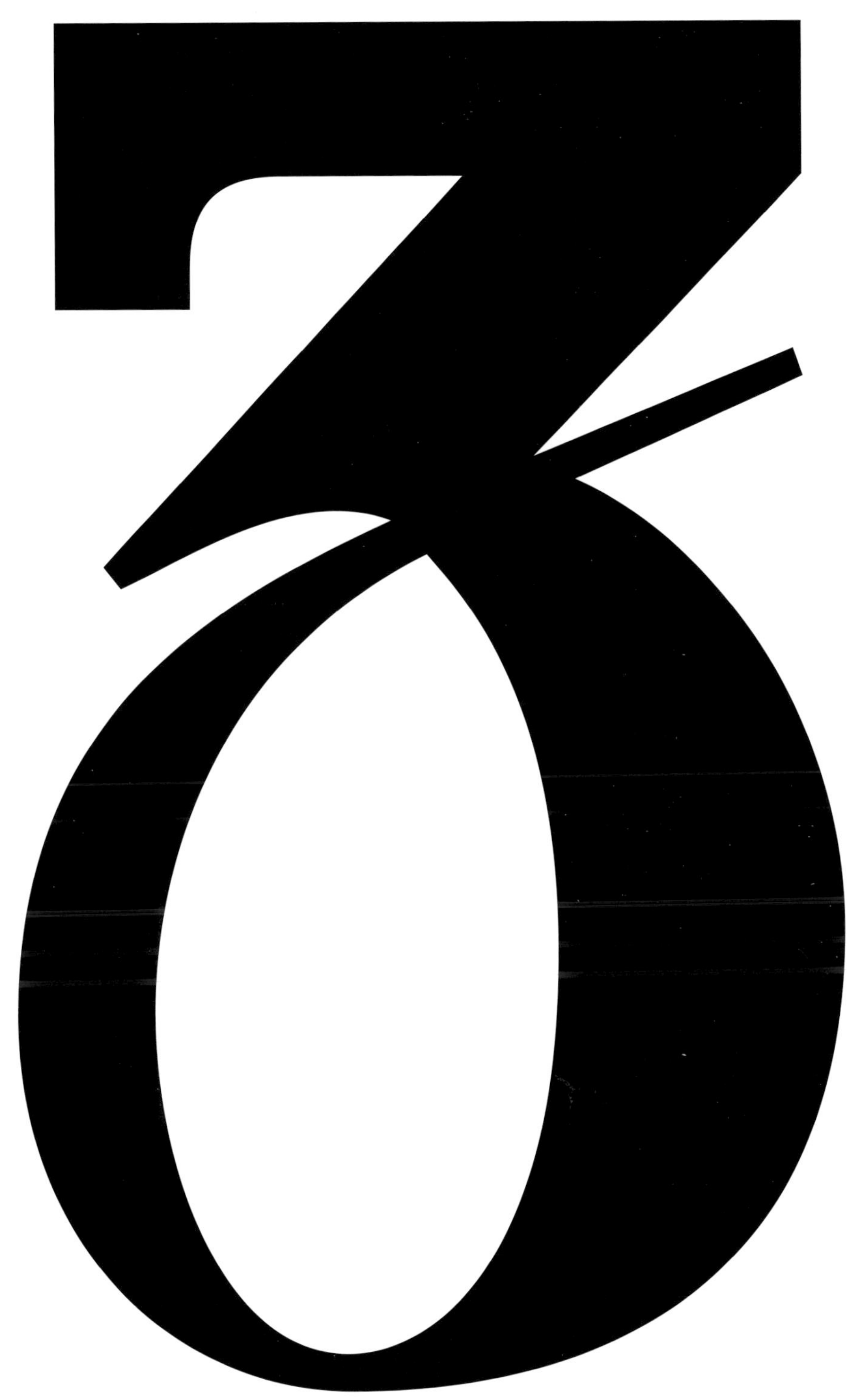

***Lars Harmsen*[1] (*1964) is Professor for Conception, Design / Typography and Layout at the University of Applied Sciences, Dortmund; he is Chief Executive of the agency »MAGMA Brand Design«[2] (Karlsruhe) as well as shareholder and Creative Director of »Melville Brand Design«[3] (Munich). In 1996 he founded, together with Uli Weiß,[4] the type foundry »Volcano Type«,[5] which at present sells more than 200 fonts. In 2004 Harmsen started the Blog »Slanted«[6] and in 2005 the magazine of the same name. Prof. Harmsen is a member of ADC e.V. and author or designer responsible of numerous publications, among them »BASTARD – Choose my Identity«, »TYPODARIUM« and »Yearbook of Type«.**

How was your interest in typography aroused? Since when have you gone in for typefaces? Did you already take an interest in them during your studies or was it afterwards?

lh: It was during my studies that I began to be interested in typography. Unfortunately, when I studied (graduation in 1996), there was not much typography training offered. At the end of my studies, I had a professor who brought the first Emigre magazines by Rudy VanderLans[7] to our courses. This was, as they say, a »mind-blowing« experience for us. This triggered my enthusiasm for typography. It became possible to create typefaces – and typographic design was exciting. The first fonts of Emigre were pixel fonts like Oakland;[8] and all this was nothing you couldn't do yourself.

Besides, I did a lot of work for the record and fashion industries during my studies. Amongst others, I worked for Fanatic, a surfboard and sportswear brand. It wasn't a big step to take from Emigre and surf riding to the Ray Gun magazine by David Carson.[9] And this is how the first fonts were created: I simply got them going. The record covers were mainly meant for techno productions. The pictures/photographs didn't play such a decisive role. Typography was more important, and Photoshop didn't exist at that time. All you could do on a computer (SE 30) was to move text from one place to another. This time had a big influence on me. You tried to achieve a lot with simple means.

You mentioned that you had the least right to be called type designer among the persons who are listed here. – What would you call yourself?

So far, I've been driven by quite another motivation when I developed my fonts. The same can certainly be said about the creation of many other fonts which are presented in this exhibition. As a matter of fact, my fonts were always meant for immediate use. For instance, there was an order for a cover, a magazine, or a T-shirt, but I couldn't find the appropriate font. It just didn't exist; it was as simple as that. That's why I created a font. I can work with typefaces rather well. However, I couldn't be a type designer because I'm too impatient. I'm more into typography.

Do you think that there are typographic trends? How do you get your information about typographic developments?

I get bored with trends. With your eyes open, you can see enough, even enough trash and »me-too« design – but at the same time enough things which are brilliant.

»Call for Type« showed many fonts reflecting the elegance of ruling pen script. A couple of years earlier, there were constructions in the sans-serif and Bauhaus style. And earlier than that, there were outstanding developments of Antiqua typefaces. Yes, there are trends. However, I wonder whether it's always necessary to jump on the bandwagon.

How does your process of designing a font look like? How do you develop type design?

The typefaces I have created so far were often conceptual. Their range of use was defined by the corresponding order. – Their realization was based on very short stages of development. Initially, I used a software called Ikarus. What a nightmare!

1 magmabranddesign.de

2 Lars Harmsen and Uli Weiß established the Karlsruher Design-Agentur Magma Brand Design in 1996.

3 The graphic design bureau Melville Brand Design was established in Munich in 2008 by Michael Schmidt, Florian Brugger, Johannes König, Lars Harmsen and Ulrich Weiß and is a partner agency of Magma Brand Design.

4 Uli Weiß, Manager of Magma Brand Design.

5 In 1996 Magma Brand Design established the German Type foundry Volcano Type, which apart from experimental display fonts also publishes large type families like the Matryoshka or the Telegramo.

6 The weblog Slanted, which started in 2004, deals with developments and trends in typography and design. Over the years, slanted.de has grown into one of the most important German-speaking design portals and discussion platforms. The magazine SLANTED, founded in 2005, is mono thematic and deals with specific subjects in typography.

7 Rudy VanderLans (*1955) is a Dutch graphic and type designer. He is married to Zuzana Licko, and together they founded the Emigre Magazine → S. 47 in 1984, a publication dealing with experimental graphic design.

8 Oakland is a series of four pixel typefaces with varying pixel sizes for low resolution display presentation. The typeface, developed in 1985 by Zuzana Licko, was published by Emigre.

9 The US typographer, type designer and surfer David Carson (*1957) became well-known for his innovative magazine design and the use of experimental typography. His style in typography and layout influenced the Grunge-aesthetics. Carson was Art Director of the magazine Ray Gun until he founded his own studio in New York in 1995 (→ S. 184).

What is your source of inspiration?

I get some inspiration from travelling, meeting other people, and interdisciplinary work. Frequently the outcome of all this is a bundle of ideas for a topic which is followed by a plan. We developed, for instance, a complete series of fonts which we called Bastard-Fonts, because we had published a book on the topic of identity. With »Bastard – Choose my Identity« we asked ourselves how something new was created in a world where so many things had already been defined and tried out. On our round-the-world trip, together with a photographer and an illustrator, we found out that basically there was quite a lot which had already been defined and that it was rather difficult to create something that was really new. However, something new could be developed by combining things and arranging them anew, as it had already been done in music for a long time. The boundaries had become blurred: Illustration, typography, graphic design, and photography – everything had been mixed; on this basis we developed the ideas for the fonts of our book.

Was there any special reason for the design of Fraktendon Pro?

Fraktendon was the first font which we had developed in the Bastard-Fonts series. We blended the American style of Clarendon[10] with a German Fraktur font. This meant reconciling softness and strictness. Fraktendon was used in the »Bastard Book«, several magazines and other printed matter.

Who are your role models?

I like people who just do things in an open-minded way, who take the lead and try something new. They don't do this because they disregard what was there before, but they want to satisfy their curiosity. Like, for example, Lars von Trier[11] whose films inspired me a lot. There was the Hollywood cinema – but his approach was quite different. David Carson also ignored conventions. He gave things a go and was very unconcerned about the criticism which he got from some conservative know-it-alls. He opened doors for generations to come. I like things which aren't based on »deeply intellectual« thought, but which rather come into being by a spontaneous impulse to try something out and find a solution. This is why these things are authentic and concrete.

Most of your projects were done in collaboration with other people like for example Boris Kahl[12] for Franktendon Pro. How important is teamwork to you?

I really like working in a team. Dialogue and exchange are inspiring and can let you know if you are on the wrong track. I find designers from other fields especially interesting. Working with people who see things quite differently can often lead to certain disagreements. These disagreements produce, however, positive results. This is why I find my work for »Slanted« so interesting. Being always in close dialogue with other creative professionals can activate the development of your own ideas. I think this way of working is much more effective than working on your own, isolated from the rest of the world, while you are waiting to hit the bull's eye. I'm not a quiet person anyway. I'm open to different opinions and not afraid of discussions. Sometimes I can be miffed about something; then I hold my tongue for a little while, think it over, and finally I might say: »You were right. Your objection was quite reasonable.« This is okay. I feel good, if I can work with other people!

The market is flooded with fonts. – Everywhere you can find free fonts of amazingly good quality. How will the demand for professionally created fonts develop in the future, and, above all, the demand for those that were created by designers?

There has been a democratization trend for quite a while now. The creation of fonts is no longer reserved for a small elite group of super specialists. We have many tools at hand, for both design and marketing. Today, actually everybody could create his own fonts, if he made an effort.

In the professional sector, the market for Non-Latin fonts will explode. There will be much movement and a lot of potential for development – also in money terms. In addition to that, there are constantly increasing standards and requirements. As designer, you don't want to do without fully developed font families. Ligatures are a must. In this field, designers' expectations have considerably increased, and free fonts cannot satisfy their hunger.

Font labels which succeed in creating overall characteristics will find a ready market. This is comparable to a well-functioning fashion label which doesn't only sells shoes, but offers at the same time appropriate tracksuit tops and caps. Especially for small studios, this can be promising. You like to buy there, because their different typefaces often go well together.

Are there any fonts you can't stand anymore?

I don't like Rotis,[13] I never did. Rotis gives me a headache.

Lars Harmsen was interviewed by Tobias Villmeter.

10 The Clarendon is a slab serif with accentuated serifs, developed in 1845 by Benjamin Fox and published in 1850 by the foundry Bauersche Schriftgießerei. In 1950 it was re-published by Hermann Eidenbenz via Linotype. It is characteristic for its drop-like endings, e.g. of the a, r or g.

11 Lars von Trier (originally Lars Trier; *1958) is a Danish film producer and scriptwriter. He is considered one of Europe's most prominent and controversial film producers. Apart from his films (e.g. Antichrist) he has also produced commercials and music videos.

12 Boris Kahl (*1975 in Schwäbisch Gmünd), studied Visual Communication at Pforzheim University. He has been Art Director of Magma Brand Design since 2001, developed numerous typefaces and is Director of the foundry Volcano Type, which was launched in 1996 (volcano-type.de).

13 Rotis is a hybrid typeface developed in 1988 by Otl Aicher → S. 156. The name was taken from Aicher's place of residence, Rotis, a part of Leutkirch/Allgäu. The special feature of this type family were the hitherto unknown weights Semi Serif and Semi Sans-Serif, as well as some unconventional single forms.

إذا الشعب يوماً أراد الحياة فلا بد أن يستجيب القدر *
ولا بد لليل أن ينجلي ولا بد للقيد أن ينكسر

Gebran2005 Bold – 40pt

واعلن في الكون أن الطموح لهيب الحياة وروح الظفر **
إذا طمحت للحياة النفوس فلا بد أن يستجيب القدر

Gebran2005 Heavy – 40pt

*
If the people one day decide to want (free) life
Then fate will surely comply

And the night will surely end
And the chains will be broken

**
And I announce to the universe
that ambition is the flame of life and the spirit of endeavor

If the souls ever strive for life
Then fate will surely comply

Schrift. Typeface. Gebran2005
Gestalter. Designer. Nadine Chahine
Label. Foundry. Monotype
Jahr. Year. 2005

»O Berg, kein Wind kann dich erschüttern« – dieses libanesische Sprichwort inspirierte die Schriftgestalterin Nadine Chahine, als sie die Gebran2005™ schuf. Die Schrift wurde entwickelt für An-Nahar, Libanons führende Tageszeitung in arabischer Sprache, und ist ein maßgeblicher Teil des neuen Designs dieser Zeitung.

Gebran2005, benannt nach Gebran Tueni, An-Nahars früherem Herausgeber, ist eine moderne Version der klassischen Titelzeile. »Die Schrift hat ein unverwechselbares, selbstsicheres Auftreten, wie Gebran selbst«, erklärt Chahine.

»Oh mountain, no wind can shake you« – this is the Lebanese proverb that inspired typeface designer Nadine Chahine as she created the Gebran2005™ typeface. Designed for An-Nahar, Lebanon's leading Arabic-language daily newspaper, Chahine's creation is part of a major redesign of the publication.

Named after Gebran Tueni, An-Nahar's former editor and publisher, Gebran2005 is a modern version of a classic newspaper headline style. »Like Gebran himself, the typeface has a distinctive, self-assured presence,« Chahine says.

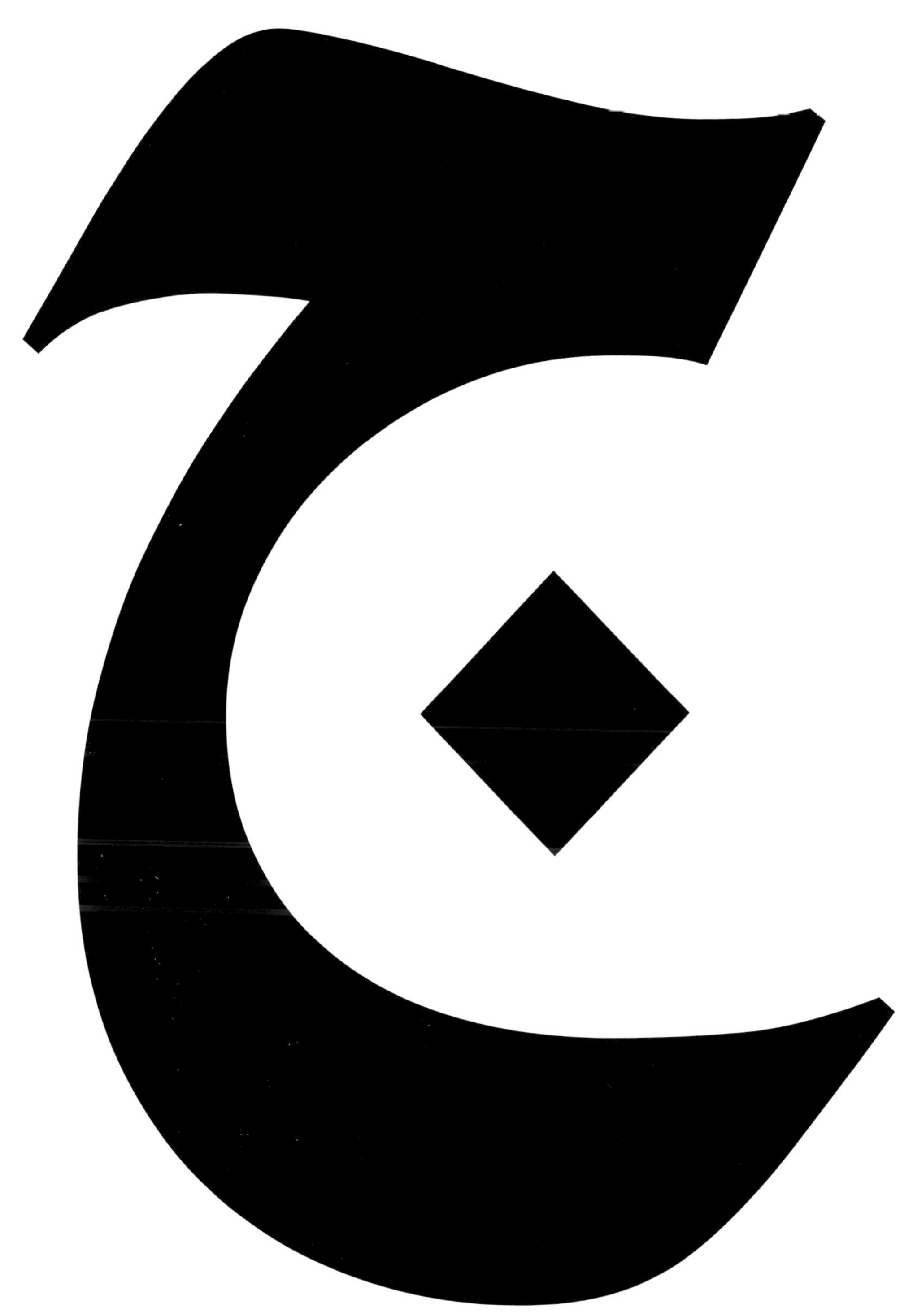

GEBRAN2005

Die libanesische Schriftgestalterin *Dr. Nadine Chahine*[1] (*1978) studierte Grafik-Design an der amerikanischen Universität Beirut. 2003 absolvierte sie ihren Master in Type Design an der Universität Reading (Großbritannien). Dort entstand Koufiya,[2] die erste Schrift, die korrespondierende arabische und lateinische Zeichen enthält. Nachdem sie arabisches Type Design an der amerikanischen Universität Dubai gelehrt hat, ist Chahine heute bei Monotype[3] Deutschland unter anderem für arabische Schriftprojekte verantwortlich. 2012 wurde Chahine über Lesbarkeitsstudien bei arabischen Schriften promoviert. Zu ihren Schriften gehören Frutiger Arabic, Neue Helvetica Arabic, Univers Next Arabic, Palatino Arabic, DIN Next Arabic, *Gebran* und Koufiya.

Woher kommt Ihr Interesse an Typografie?

nc: Zu Beginn meines Grafik-Design-Studiums hatte ich einen wunderbaren Lehrer, Prof. Samir Sayegh, der Kalligraf, Kunstkritiker und Kunsthistoriker ist. Das war ein großer Glücksfall für mich. Er unterrichtete arabische Typografie, und wir sollten mit einer Schrift, die er entworfen hatte, unsere Namen schreiben. Seine Schriftzeichen waren sehr ornamental und hatten viele Verzierungen, was ich in der Typografie gar nicht mag. Also fragte ich ihn, ob ich die Verzierungen bei den Endungen weglassen kann, und er war einverstanden. Ich verbrachte zwei Wochen damit, hin und her zu ändern, und als ich es schließlich so hatte, wie ich es wollte, da hatte es mich gepackt.

Natürlich war es nur eine Übung im Buchstabenzeichnen, kein Schriftentwurf. Aber trotzdem hat es mir die Augen geöffnet. Damals waren die arabischen Fonts, die wir hatten, schlecht gezeichnet und ziemlich hässlich, und da zeigte er uns jetzt diese erstaunliche Kalligrafie. Die Diskrepanz gegenüber dem, was wir bisher kannten, war schockierend, und es war wie eine Aufforderung zum Handeln. In meinem dritten Studienjahr hatte ich dann eine weitere Klasse über arabische Schrift und fing an, mich im Rahmen meiner Projekte zu Verpackung und Posterdesign mehr mit arabischen Schriftzeichen zu befassen, wobei ich versuchte, immer mehr davon in meiner Arbeit unterzubringen. Und schließlich war mir klar, dass es das war, was ich machen wollte. Es war ein wirklicher Glücksfall, dass ich den Schwerpunkt für meine Arbeit so früh in meiner Studienzeit fand.

1 arabictype.com

2 2003 gestaltete Nadine Chahine die Schrift Koufiya im Rahmen ihres MA-Projekts in Reading (UK). Es ist die erste Schrift, bei welcher der arabische und der lateinische Zeichensatz im selben Zeitraum von Hand von einem einzelnen Gestalter gezeichnet wurde. 2012 wurde Koufiya von Linotype veröffentlicht.

3 Die 1890 gegründete Monotype GmbH vertreibt bzw. lizensiert heute hochwertige digitale Schriften über monotype.com Außerdem entwickelt Monotype Corporate Fonts für Geschäftskunden und stellte 2005 das Font Management-Programm Font Explorer zur Verfügung (→ S. 51, 60).

Wie lange brauchen Sie, um einen Font zu entwickeln? Wann sind Sie fertig damit?

Meine erste richtige Schrift entwickelte ich während meines Masterprogramms an der Universität von Reading. Daran habe ich ein ganzes Jahr gearbeitet, und einige Jahre später hatte ich 150 Versionen dieser Schrift, es war schon verrückt. Aber es war ein Lernprozess. Man lernt, Kurven zu zeichnen, man lernt, über Abstände zu entscheiden, man versucht, die Beziehungen zwischen zwei Schriften herzustellen, was niemand zuvor gemacht hatte, das war ein sehr komplexes Projekt.

Heutzutage entwerfe ich vielleicht 15 bis 20 Varianten einer Schrift, ehe ich die endgültige Version habe. Es hängt natürlich auch davon ab, ob man zeitlich unter Druck ist; wenn es für einen Auftraggeber ist, dann ist es meist dringend. Dann dauert es vielleicht einen oder zwei Monate. Wenn es ein Projekt für die Bibliothek ist und man oft unterbrochen wird, kann es über ein Jahr dauern.

Wenn man sich mit Schriftdesign befasst, ist der Anfang immer sehr schwer, denn man weiß nicht, was man eigentlich will und es ist recht schwierig, bis die Buchstaben wie wirkliche Buchstaben aussehen, weil man die Proportionen nicht erkennt. Es ist, als wollte man aus dem Nichts eine menschliche Gestalt zeichnen. Man sieht einen Menschen an und versucht ihn zu zeichnen, aber man ist sich über die Proportionen nicht im Klaren. Man zeichnet den Kopf, die Füße, den Körper und die Arme, aber irgendwie wirkt das alles nicht sehr überzeugend. Aber wenn man dann anfängt zu verstehen, wie die einzelnen Elemente zueinander in Beziehung stehen und es plötzlich doch wie ein Mensch aussieht, dann hat man dieses Aha-Erlebnis. Dann muss man irgendwann aufhören zu zeichnen, damit der Mensch laufen lernt. Und plötzlich sieht man ihn auf Reklametafeln.

Ihr erster Font war Koufiya. Es war der erste Font mit übereinstimmenden arabischen und lateinischen Schriftzeichen. Wie schafft man es, zwei so verschiedene Schriften einander anzupassen?

Als ich in Reading damit anfing, wusste ich, dass ich eine lateinische und eine arabische Schrift zeichnen wollte. Anfangs war ich ganz von dem Problem in Anspruch genommen, wie ich die Endungen gestalten sollte. Ob ich Serifen wollte oder nicht, und welche Art von Serifen. Als ich dann die lateinische und die arabische Version fertig hatte und sie nebeneinander legte, passten sie nicht wirklich zusammen.
Da bekamen wir Besuch von einem weiteren ganz phantastischen Schriftdesigner: Jean François Porchez.

Als er dazu kam, hatte ich sieben Versionen der lateinischen Schrift, die ich an die arabische anpassen wollte. Er riet mir, mich auf die Strukturen zu konzentrieren und nicht auf die Serifen. Er fand, dass meine arabische Schrift viel moderner aussah als meine lateinische, und er riet mir, auf die Kurven in den Strukturen zu achten, um sie besser einander anzupassen.

Und plötzlich ging mir ein Licht auf. Es ist nicht das Dekorative, worauf es ankommt. Wenn zwei Menschen äußerlich miteinander harmonieren sollen, dann hat das nichts mit ihren Schuhen zu tun, sondern mit ihrem Körperbau, erst dann treten sie in Bezug zueinander. Man muss darauf achten, wie die Bewegungen zueinander passen. Sind sie dynamisch? Oder entspannt? Sind sie offiziell? Denn es liegt in der Bewegung, die den Buchstaben entstehen lässt. Welchen Rhythmus habe ich? Ist er eng? Oder schnell? Ist er langsam, oder eher geometrisch? Ist er organisch? Ist er steif oder flüssig? Man sieht sich die Körper an, die Skelette der beiden. Und auf diese Weise schafft man eine Beziehung zwischen ihnen. Dann versucht man, eine gewisse Harmonie herzustellen. Also hängt es nicht von der Verzierung ab, sondern vom Skelett. Das zeichnet man erst, dann gibt man Fleisch auf die Knochen. Und dann fängt es an, sich zu bewegen. Mit dieser Bewegung arbeitet man und so schafft man es, dass sie harmonieren.

Es ist aber nicht so, dass aus dem Mann eine Frau wird, oder aus der Frau ein Mann. Denn sie haben ja einen grundsätzlich verschiedenen Körperbau. Frauen haben breitere Hüften, Männer breitere Schultern. Sie sind verschieden proportioniert und man kann das eine nicht in das andere verwandeln. Also muss das Arabische arabisch bleiben, und das Lateinische bleibt lateinisch. Man muss nur für Harmonie zwischen beiden sorgen, damit sie nebeneinander existieren können.

Aber wie überträgt man lateinische Schriftzeichen in arabische Schriftzeichen? Zum Beispiel Helvetica[4] Arabic oder Zapfino[5] Arabic. Wie funktioniert das?

Für die Übertragung ins Arabische arbeiten wir eher mit einer Form von Verwandschaft. Wir sehen uns die lateinische Schrift an und versuchen zu verstehen, welche Charakteristika sie mitbringt. Ein Beispiel ist Frutiger[6] Arabic. Zunächst ist das eine humanistische Sans Serif, an der man die Bewegung der Feder erkennt. Die Schrift ist freundlich, aber auch nicht zu informell. Aber sie ist auch nicht so steif wie die Helvetica oder eine Grotesk. Außerdem ist sie eine Schrift zur Beschilderung, denn ursprünglich war sie für den Flughafen Charles de Gaulle in Paris gedacht. Man übernimmt das Konzept und auch etwas von der visuellen Gestaltung, statt das ganze Schriftbild zu ändern. Man arbeitet daran, bis sie ähnlich genug sind, um nebeneinander bestehen zu können, aber man kann nicht das Eine in das Andere verwandeln.

Können Sie uns etwas über die Palatino[7] Arabic erzählen?

Ich arbeitete mit Prof. Hermann Zapf[8] an seiner Al-Ahram-Schrift,[9] die er in den 1950er Jahren für eine Zeitung in Kairo entwickelt hatte. Es war eine gute Zusammenarbeit, denn ich habe viel von ihm gelernt. Zum Beispiel wie er zeichnet, wie er die Dinge sieht, er hat unglaublich scharfe Augen. Er ist ein erstaunlicher und wirklich legendärer Schriftdesigner. Für mich war es fast wie ein zweiter Masterkurs, einfach mit ihm zusammenzuarbeiten und zu lernen, Schrift mit seinen Augen zu sehen. Interessanterweise entwarfen wir die Schrift für ein klassisches Buch. Wenn man die lateinische und die arabische Schrift vergleicht, wird man merken, dass das Arabische viel organischer, flüssiger und dadurch dynamischer ist als das Lateinische. In diesem Fall mussten wir uns bei der Frutiger Arabic zuerst mit der Funktion befassen.

Es ist leichter, eine Korrespondenz zu einer Sans Serif zu entwerfen, weil man hier die Funktion und die ästhetischen Anforderungen gleichzeitig berücksichtigen kann. Wenn man dagegen eine Korrespondenz zu einer Serifenschrift entwirft, muss man sich auf die Funktion konzentrieren, denn im Arabischen gibt es keine Serifen, und wenn man ein traditionelles Textbild auf Arabisch anstrebt, gibt es bestimmte Anforderungen, die erfüllt werden müssen. Das rückt es visuell vom Lateinischen weg, was man aber akzeptieren muss. Wenn man also ein traditionelles Schriftbild für ein Buch anstrebt, muss man berücksichtigen, wie das aussehen muss, damit die Leute es lesen wollen. Dasselbe gilt für die Beschilderung, wo es ein glücklicher Zufall ist, wenn wir es mit einer Sans Serif zu tun haben, weil es dann leichter ist, gleichzeitig die ästhetischen und die funktionellen Anforderungen zu erfüllen. Also wird man sehen, dass das Arabische dynamischer ist, aber das erwartet man ja auch von einer Buchschrift.

Ein interessantes Detail ist folgendes: Wir entwarfen also die Schrift, und irgendwann dachten wir, wir seien fertig. Nun gab es aber einen Meister der arabischen Kalligrafie, Ibn Muqla, der vor 900 Jahren lebte und der bestimmte Proportionen festlegte, nach denen Schriftkünstler sich richten sollten. Ich prüfte nach, ob mein Entwurf mit diesen Proportionen übereinstimmte, und das war nicht der Fall. Also änderte ich meinen Entwurf entsprechend ab, und plötzlich wirkte es viel besser. Das System der Proportionen, das vor Hunderten von Jahren für uns entwickelt wurde, gilt also noch heute. Was für ein Glück für uns!

4 Die Schriftfamilie Helvetica ist eine serifenlose Linear-Antiqua mit klassizistischem Charakter und gehört zu den am weitesten verbreiteten Groteskschriften. Die ersten Schriftschnitte wurden ab 1956 von Max Miedinger und Eduard Hoffmann gestaltet (→ S. 51, 183).

5 Zapfino ist eine Schrift, die von Hermann Zapf entworfen wurde. Die Schriftfamilie Zapfino umfasst insgesamt 18 Schriftschnitte und ist im Sinne der DIN 16518 als eine Schreibschrift klassifiziert. Zapfino wurde 1999 mit dem Designpreis des Type Designers Club ausgezeichnet.

6 Frutiger ist eine serifenlose Linear-Antiqua-Schrift, die 1975 von Adrian Frutiger entworfen und von der Schriftgießerei D. Stempel veröffentlicht wurde (→ S. 51).

7 Die Schriftfamilie Palatino ist eine französische Renaissance-Antiqua, die 1986 von Hermann Zapf für die D. Stempel AG in Frankfurt am Main entworfen wurde. Im Jahr 1999 überarbeitete Linotype die Palatino und erweiterte die Familie um den lateinischen, griechischen und kyrillischen Zeichensatz.

8 Hermann Zapf (*1918) ist Typograf, Kalligraf und Autor. Er hatte u.a. Lehraufträge in Darmstadt und Rochester und entwarf über 200 Schriften, u.a. Palatino, Optima, und Zapfino.

9 Al-Ahram ist eine 1956 von Herman Zapf gestaltete Schrift für die Al-Ahram-Zeitung in Kairo. 2005 wurde sie überarbeitet und in die Palatino Nova-Familie eingegliedert.

الخط العربيِ يبقى عربياً واللاتيني لانينياً. نحاول أن نخلق حواراً بينهما لكيِ يستطيعا العيش سوياً.

GEBRAN2005

»So the Arabic needs to stay Arabic.
The Latin needs to stay Latin.
You just create harmony between them,
so that they can coexist.«

Nadine Chahine
→ S. 87

Hermann Zapf ist nicht der Einzige, der eine Schrift für eine Zeitung entwickelt hat. Sie selbst haben Gebran für die libanesische Zeitung An-Nahar entworfen. Wie fühlt man sich, wenn man seinen Font auf der größten Tageszeitung des Libanon sieht?

Das ist schon toll. Dies war vielleicht die persönlichste Arbeit, die ich jemals für einen Kunden gemacht habe. Ich wuchs auf mit dieser Zeitung, die mein Vater jeden Tag nach dem Mittagessen nach Hause brachte. Er schreibt noch immer ab und zu für sie. Es war unglaublich, es war das erste große Projekt im Libanon. Ich hatte davor schon viele andere Projekte, aber keines davon war für den Libanon, also nicht für meine Heimat. Aber diese Zeitung habe ich zu Hause immer gelesen, deshalb gab es da diese persönliche Beziehung. Außerdem hatte es mit der allgemeinen Situation im Libanon zu tun, denn der Chefredakteur und einer der besten Journalisten fielen einem Attentat zum Opfer. Und dieses neue Design der Zeitung war eine Aussage, es sollte zeigen, dass diese Zeitung weiter bestehen würde, trotz aller Bedrohungen, aller Attentate und anderer Tragödien.

Wenn man sich fragt, in welcher Art von Land man leben möchte, bedeutet der Begriff Pressefreiheit in diesem Zusammenhang plötzlich so viel mehr. Das Design für die zwei Schriften war in drei Wochen fertig, es war der reine Wahnsinn. Ich habe gearbeitet wie besessen, aber man entwickelt eine Motivation und eine Energie, wie man sonst eher selten hat. Dies war eine ungeheuer befriedigende Arbeit, das Design ist sehr kraftvoll. Wenn ich nach Hause komme und die Zeitung sehe, freue ich mich immer, denn auf diese Weise kann ich visuell einen Beitrag zu meiner persönlichen Umgebung leisten.

Sie sprechen von einem persönlichen Anliegen. In einem anderen Interview sagten Sie auch, dass die Alphabetisierungsrate in Nahost immer noch sehr niedrig ist, und dass es die beste Art und Weise sei, etwas zu verändern, wenn man die Menschen mehr zum Lesen bringt. Glauben Sie, dass die Typografie helfen kann, diese Probleme zu lösen?

Ja, davon bin ich überzeugt. Natürlich kann man mit Typografie das Lesen beeinflussen. Zum Beispiel kann man eine Zeitung oder ein Buch in die Hand nehmen und die Schrift macht einem Kopfschmerzen, weil sie schlecht gesetzt und wenig attraktiv ist. Oder man schreibt in einem zu gewundenen Stil, was die Leute auch abschreckt.

Es gibt viele entscheidende Zusammenhänge zwischen dem Design und seiner Auswirkung auf das Lesen. Denn was wir schließlich lesen, sind ja die Schriftzeichen, und die Fähigkeit, diese zu erkennen, macht das Lesen für uns leichter oder schwerer. Deshalb ging es in meiner Doktorarbeit auch darum, zu untersuchen, welche Auswirkung die Komplexität des Stils auf Lesbarkeit und Lesegewohnheiten hat. Je komplexer der Stil, desto schwieriger ist er zu lesen.

Das Interview mit Nadine Chahine führten Anna Alexander, Lisa Grünwald und Bahar Hasan.

The Libanese type designer *Dr Nadine Chahine*[1] (*1978) studied graphic design at the American University of Beirut. In 2003, she finished her master's degree in type design at the University of Reading (Great Britain). This is where Koufiya[2] was created. Koufiya was the first typeface to contain both Arabic and Latin designed simultaneously to correspond to each other. After teaching at the American University of Dubai, Chahine is now, among other things, responsible for Arabic type projects of Monotype[3] Germany. In 2012, Chahine did a doctor's degree with a thesis concerning legibility studies in the field of Arabic writing. Among her fonts are Frutiger Arabic, Neue Helvetica Arabic, Univers Next Arabic, Palatino Arabic, DIN Next Arabic, *Gebran2005* and Koufiya.

Where does your interest in typography come from?

nc: It started in my graphic design class, where I had the great good luck to have an amazing teacher, Prof. Samir Sayegh. He's a calligrapher, art critique and historian. He gave us an Arabic typography class where we were supposed to take a typeface he had designed and build up our names. His design was very ornamental and with many flourishes, which I don't like in typography. I asked him if I could remove the flourishes from the terminals and he agreed. I spent two weeks going back and forth, making changes and eventually I got it to where I wanted it to be and it sort of gave me the design bug.

It was a lettering exercise obviously, not a typeface. But still, it opened my eyes. At the time, the actual Arabic fonts we had were so badly drawn and ugly, but then he showed us amazing calligraphy. The discrepancy between the two was shocking. That was like a call to action. In my third year I had another course about arabesque and I started doing a lot of Arabic lettering for my projects in packaging and poster design, trying to get more and more Arabic into my work. Then I knew, this was what I was going to do. It was a great stroke of luck that I found my place so early on at university.

How long does it take you to design a font? When is it finished?

My first proper typeface was the one I did during my MA at Reading. I spent a whole year working on that one, and a few years later I had 150 versions of the typeface. It was crazy. But it was a learning process. You're learning how to draw curves, you're learning how to space, you're trying to address the relationship of the two scripts, which nobody had done before. So it was a very complex project.

These days I think I go through maybe 15–20 versions of a typeface before I arrive at the final one. The time spent depends on how complex it is, how quick we have to be, whether it's for a client and we're in a rush. It could take one month or two. If it's a project for the library and I keep getting interrupted, it could take more than a year.

When you get into typeface design, the beginning stages are so hard, because you don't know what you're doing and it's≈so hard to get the letters looking like actual letters, because your eyes don't see the proportions. It's sort of as if you decide to draw a human figure from scratch. You look at humans and you try to draw them, but you haven't nailed down the proportions. You draw the head, you draw the feet, you draw the body, the arms, but somehow it doesn't look very convincing. When you start to understand how the elements relate to one another and it suddenly starts to look like a person, you have this aha-moment. At some point, youneed to stop because this person needs to start walking. And then you start seeing the person on poster billboards.

The first font you did at Reading was Koufiya. It was the first font with matching Arabic and Latin letters. How is it possible to match two such completely different scripts?

When I started the exercise at Reading, I knew I wanted to draw a Latin and an Arabic version. My entire thinking was about the terminal treatments at first. Whether to give it serifs or not and then what kind of serifs. After doing the Latin and the Arabic I put them next to each other and things were not really working. Then we had a visit from another amazing typeface designer: Jean François Porchez.

He came in and by then I had seven versions of a Latin that could match with an Arabic that I wanted to design. He told me to look at the structure, not at the serifs. He thought my Arabic looked much younger than my Latin and that I needed to look at the curves of the structure to find the combination.

And suddenly the bulb started flashing in my head. What you need to look for is not the decoration. If you want to have two people that are in harmony with each other, it's not about the shoes they are wearing, it's about the build. That's how you get them to talk to each other. You need to see what kind of movement you have. Is it dynamic? Is it relaxed? Is it formal? The kind of movement which draws the letter. What kind of rhythm do you have? Is it tight? Is it fast? Is it slow? Is it very geometric? Is it very organic? Is it stiff? Is it fluid? You look at the bodies, the skeletons of the two. That's how you start to create a relationship. Then you try to find a harmony between them. If you go back to the human body we're drawing, it's not about the face or the teeth, it's about the skeleton. First you draw the skeleton, then you put the meat on the bones. And then it starts to move. You work with the movement and get them to work together.

You always need to make sure that you're not trying to turn one into the other. The man doesn't become a woman.

1 arabictype.com

2 Nadine Chahine developed her typeface Koufiya in 2003 as part of her Master project at Reading University. It is the first typeface where the Arabic and the Latin characters were drawn by one designer and within the same timeframe. Koufiya was published in 2012 by Linotype.

3 Monotype GmbH, established in 1890, sells and licences high quality digital fonts via monotype.com. Monotype also develops corporate fonts for clients and in 2005 made the Font Management Programme Font Explorer available (→ S. 54, 63).

اذا اردت ان تجد توازن بين شخصين مختلفين فالحل ليس في تلاؤم الأحذية وأما في تلاؤم الهيكل الأساسي.

»If you want to have two people that are in harmony with each other,
it's not about the shoes they are wearing,
it's about the build.«

Nadine Chahine

→ S. 85

The woman doesn't become a man. Because they have inherently different structures. Women will have wider hips, men will have wider shoulders. They will have different proportions. You cannot morph one into another. So the Arabic needs to stay Arabic. The Latin needs to stay Latin. You just create harmony between them, so that they can coexist.

But how do you transfer Latin types into Arabic types? Like Helvetica[4] Arabic or Zapfino[5] Arabic. How does that work?

In Arabic, we basically make a companion. We look at the Latin and then we try to understand what sort of characteristics this Latin has. For example, the Frutiger[6] Arabic. First, it's a humanist sans-serif. You see some of the influence of the pen's movement. It's very friendly, but not too informal. But it is not as stiff as a grotesque. And then it's designed for signage, because it was initially meant for Charles de Gaulle Airport in Paris. You adopt the concept and some of the visual treatment, rather than transforming the typeface. You work on it until they work together, rather than trying to turn them into the same entity.

Can you tell us something about Palatino[7] Arabic?

I started working with Prof. Hermann Zapf[8] on his Al-Ahram[9] typeface, which is a typeface he designed for a newspaper in Cairo in the Fifties. It was a really nice collaborative effort because I learned so much from him. The way he draws, the way he looks at things, how sharp his eyes are. He's an amazing and legendary designer. It was almost like a second Master Degree just being able to work with him and seeing typefaces through his eyes. The interesting thing is, we designed it for a classic bookface. When you put the Latin and the Arabic next to each other, you will notice that the Arabic is much more organic, much more fluid and much more dynamic than the Latin. In that case, we had to consider first the function just as we did with the Frutiger Arabic. It's easier to design a companion to a sans-serif because you can fulfill both, the function and the aesthetic requirements, at the same time. But when you're designing a companion to a serif typeface, you have to choose function, because we don't have serifs in Arabic and if you go for a traditional textface there are certain visual requirements, and that takes it away from the Latin and you have to accept it. When it comes to a traditional bookface you need to consider what a traditional bookface looks like, so that people will want to read it. With signage it is the same, it is a happy coincidence that with a sans-serif it's easier to fulfill both, the aesthetics and the functional requirement. So you will find that the Arabic has more dynamic energy to it, but that's what is to be expected in that style of book typeface.

The interesting thing was, we were designing it and at some point we thought we were finished. However, a long time ago there was an Arabic calligrapher, Ibn Muqla, who laid down a system of proportions for Arabic letterforms. So I checked if my design conformed with this system, and it did not, but after I had changed it accordingly it looked so much better. The system that was developed for us hundreds and hundreds of years ago still gives us the ideal set of proportions, a very lucky situation for us.

Hermann Zapf is not the only one who developed a typeface for a newspaper. You designed Gebran2005 for the Lebanese newspaper An-Nahar. How does it feel, seeing your font on the biggest newspaper of Lebanon?

It's amazing. That was probably the most personal of all projects I've done for clients. I grew up reading the newspaper, as my dad brought it home every day after lunch. My dad still writes for the newspaper every once in a while. It was amazing because it was the first big project in Lebanon. I had done so many other projects, but none of them were for Lebanon, so it's not home. This is the newspaper I read in my actual home and so there was of course this aspect to it. And then the other aspect was related to the general situation in Lebanon, because the chief editor and one of the best journalists were assassinated. The redesign of the newspaper was a statement, showing that it would continue in spite of all the intimidation and political assassinations and the tragedies surrounding it.

So there is a much wider aspect to it when you talk about the freedom of the press and what kind of country you want to have. The design of the two typefaces was finished in three weeks, which is much faster than usual. I was working around the clock but it's a kind of energy and motivation that you don't find often in typical projects. It was one of my most fulfilling projects. The design is very powerful. When I go home and the newspaper is there, it's nice, because it is a way of visually contributing something to my personal environment.

You're talking about your personal mission. You also said in another interview that the Middle East has a very low literacy rate and the best way to solve these problems is to encourage people to read more. Do you think that typography can help to solve these kinds of political issues?

Yes, absolutely. Typography can definitely affect reading, that is for sure. Because you can pick up a newspaper or a book and it gives you a headache because the typeface may be badly set and is not inviting. Or the style of writing is too complex and it puts you off.

There is a strong relationship between the design and the effect on reading. Because what we are reading in the end are the symbols, and the ability to recognize them makes the reading process easier or more difficult. So my PhD project was specifically to look at the effect which legibility has on reading. The more complex the style, the more difficult it is to read.

Nadine Chahine was interviewed by Anna Alexander, Lisa Grünwald and Bahar Hasan.

4 The Helvetica type family is a sans-serif Linear-Antiqua with a classicistic character and one of the most frequently used Groteques. The first styles were designed from 1956 onwards by Max Miedinger and Eduard Hoffmann (→ S. 54, 185).

5 Zapfino is a typeface developed by Hermann Zapf. The type family Zapfino comprises a total of 18 styles and is classified, according to DIN 16518 as a calligraphic typeface. Zapfino was awarded the Design Prize of the Type Designers Club in 1999.

6 Frutiger is a sans-serif Linear-Antiqua typeface, created in 1975 by Adrian Frutiger and published by the type foundry D. Stempel (→ S. 54).

7 The type family Palatino is a French Renaissance Antiqua developed in 1986 by Hermann Zapf for D. Stempel AG in Frankfurt am Main. In 1999 Linotype reworked the Palatino and enlarged the family by a Latin, Greek and Cyrillic character set.

8 Hermann Zapf (*1918) is a typographer, calligrapher and author. He taught in Darmstadt and Rochester, among others, and has designed more than 200 typefaces, among them the Palatino, Optima and the Zapfino.

9 Al-Ahram is a typeface designed in 1956 by Hermann Zapf for the Al-Ahram Newspaper in Cairo. In 2005 this was revised and incorporated into the Palatino Nova family.

Grow BCEF

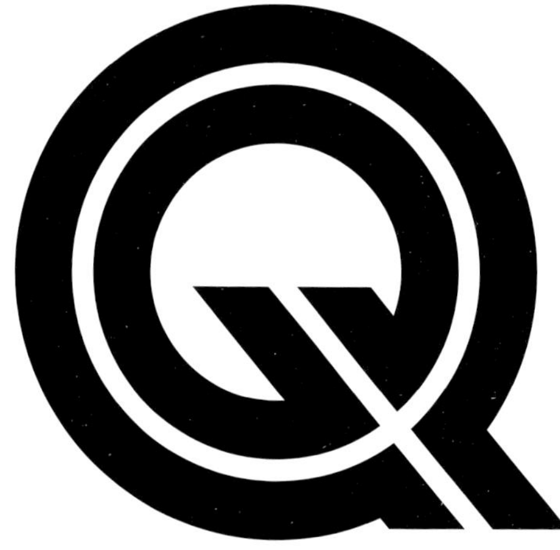

Grow AF

Grow BCE

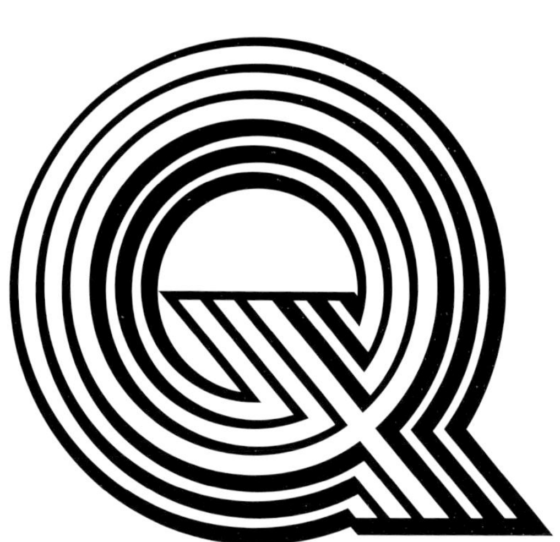

Grow ABCDEF

Schrift. Typeface. Grow
Gestalter. Designer. Johannes Breyer & Fabian Harb
Label. Foundry. dinamo.us
Jahr. Year. 2013

Die ursprüngliche Anregung für Grow kommt von den bekannten Lettera-Büchern von Niggli, in denen verschiedene Mehr-Schichten-Schriften gezeigt werden. Diese geschichteten Schriftformen erweckten das Interesse für Schriften mit mehreren Ebenen. Mit Hilfe moderner Software, die zum Entwickeln von Schriften eingesetzt werden kann, wurden die einzelnen Schichten auseinander genommen, um sie separat zu verwenden.

Aus technischer Sicht verfügt Grow über einen kompletten Latin Extended Zeichensatz, was für Display-Fonts eher selten ist. Mithilfe des Schriften-Ingenieurs Gustavo Ferreira und seinen maßgeschneiderten Anwendungen wurden 63 mögliche Formenkombinationen von Grow entwickelt. Die Schnitte von Grow können alle einzeln kombiniert, eingefärbt und verschoben werden, was unzählige Variationen ermöglicht.

Grow's initial inspiration derive from the famous Niggli's Lettera books in which various multi-level typefaces are shown. Discovering those layered letterforms, a typeface with multiple layers was developed. Making use of the possibilities which modern software offers in the design process, the layers were separated in order to use each of them singly.

On the technical side, Grow comes up with the support of a complete Latin Extended character set, rather rare for display fonts. With the help of type engineer Gustavo Ferreira and his custom tailored applications 63 possible combinations of Grow's shapes were created. Grow's cuts can all be combined, coloured and shifted individually, creating a multitude of variations.

IMPROVE

PEAK

LUXURY

OPTIMUM

MORE

JACKPOT

Grow A
Grow EF
Grow ABEF

Grow ACDF
Grow BDE
Grow CDF

Johannes Breyer[1] (*1987) stammt aus Erlangen, Deutschland. Nach dem Studienbeginn an der ZHdK[2] und einem Praktikum bei »Norm«,[3] beendete er 2013 sein Studium an der Gerrit Rietveld Akademie in Amsterdam.

Fabian Harb[1] (*1988) kommt aus St. Gallen, Schweiz, und absolvierte die Typografie-Klasse der Schule für Gestaltung in Basel. Er arbeitet zusammen mit Laurenz Brunner[4] sowie an eigenen Arbeiten. Johannes Breyer und Fabian Harb betreiben gemeinsam die Type Foundry »Dinamo«.[1]

Wie habt ihr euch kennengelernt und welche Projekte habt ihr schon zusammen gemacht? Wie funktioniert eure Zusammenarbeit?

jb+fh: Wir haben uns durch Larissa,[5] eine gemeinsame Freundin, während des Studiums in der Schweiz kennengelernt. Daraus entstand ein Austausch zwischen Basel und Zürich, der sich durch unsere ganze Studienzeit zog. Nach dem Abschluss zog Fabian für ein Praktikum nach Amsterdam, während Johannes dort sein Studium fortsetzte. In der Schweiz wie auch in Holland arbeiteten wir immer wieder an gemeinsamen Projekten und Schriften. Mittlerweile leben und arbeiten wir beide in Berlin.

Seht ihr euch als Schriftgestalter oder eher als Grafik-Designer, die ihr eigenes Werkzeug herstellen?

Wir verstehen Grafik als einen Dialog zwischen allen beteiligten Instanzen. Dies hat grundlegend mit Sprache zu tun, Sprache wiederum mit Schrift und Schrift mit Form. So scheint es uns sinnvoll, schon diese kleinsten Bestandteile dem Auftrag entsprechend zu formen.
Wir beginnen viele unserer Arbeiten mit dem Entwurf von Schriften, doch um die Frage zu beantworten: Wir sind beide gelernte Grafiker – uns Schriftgestalter zu nennen wäre eine etwas wilde Behauptung.

Johannes Breyer, Du machst gegenwärtig Deinen Abschluss an der Gerrit Rietveld Akademie in Amsterdam. Was fasziniert Dich an der Rietveld und hat Dich zu einem Studium dort bewogen? Wie unterscheiden sich die Mentalität und die Art der Gestaltung dort von Schweizer Hochschulen und Agenturen?

jb: Am meisten profitiere ich vom angeregten Austausch unter den Lehrern und Schülern, von denen die meisten von überall her für eine kurze, intensive Zeit an diese Schule pilgern. Diese bunte Mischung aus verschiedenen Arbeits- und Sichtweisen stellt viele Fragen und schafft Raum,

1 johannesbreyer.com
fabianharb.ch
dinamo.us

2 Die Züricher Hochschule der Künste (ZHdK) entstand 2007 aus der Fusion der Hochschule für Gestaltung und Kunst (HGKZ) und der Hochschule für Musik und Theater (HMT). Das Museum für Gestaltung Zürich ist der ZHdK angegliedert (→ S. 22).

3 »Norm« wurde 1999 von Dimitri Bruni und Manuel Krebs gegründet. Zu den Kunden zählen Omega, Swatch und der Louvre. Manuel Krebs und Dimitri Bruni, die für ihren ausgesprochen systematischen Ansatz in der Typografie bekannt sind, haben über die Type Foundry Lineto sehr erfolgreiche Schriften veröffentlicht (u. a. LL Simple, LL Replica).

4 Nach seinen Studien in London (Central Saint Martins) und Amsterdam (Gerrit Rietveld Academie) lebt und arbeitet Laurenz Brunner (*1980) heute in Amsterdam, Berlin und Zürich. Bei Lineto veröffentlichte er die LL Akkurat, 2013 erschien seine LL Circular.

5 Larissa Kasper studierte Visuelle Kommunikation an der ZHdK. Zusammen mit Rosario Florio gründete sie das Design-Studio Kasper-Florio in St. Gallen.

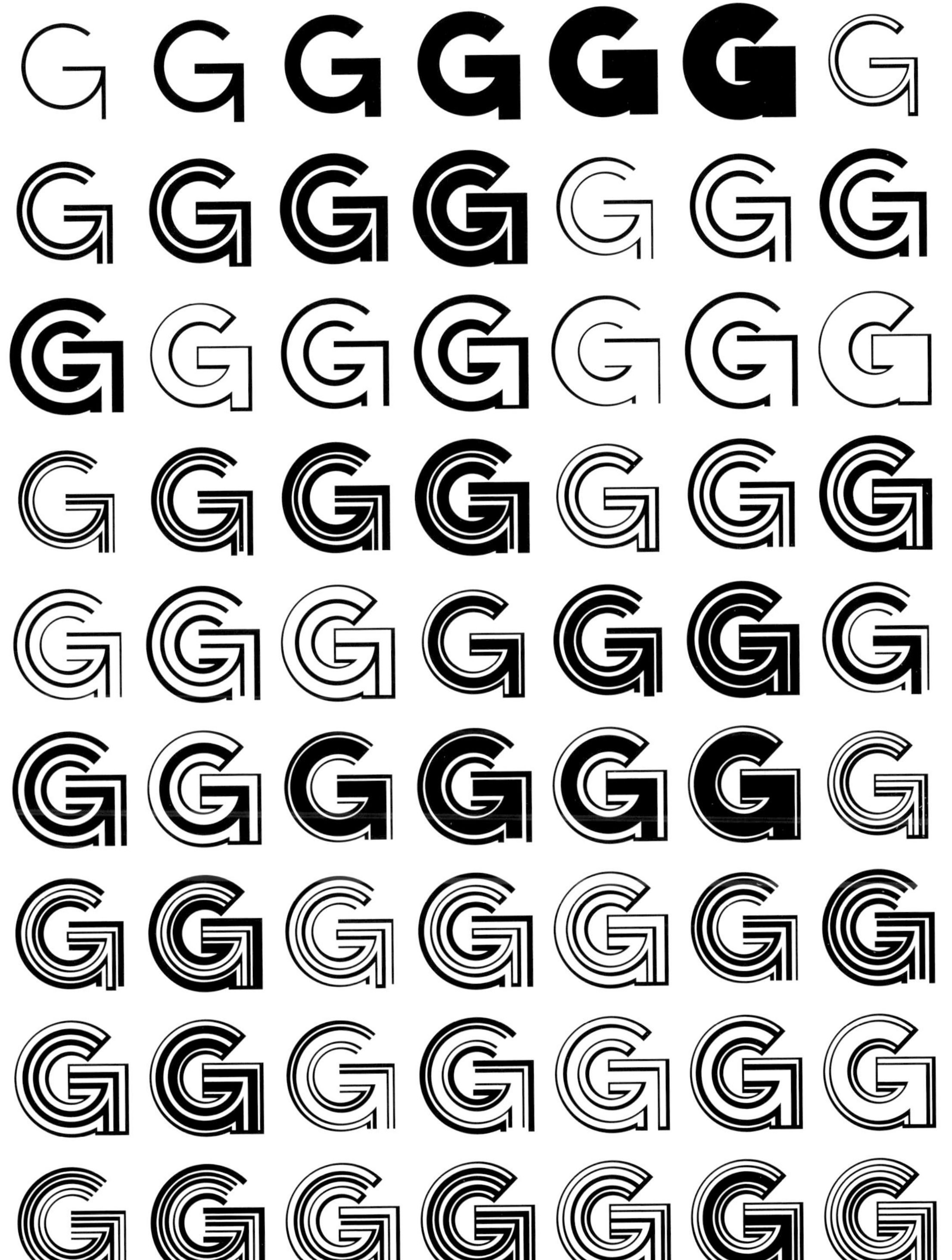
GROW

sich diesen auf unterschiedliche Weisen – und unterschiedlich erfolgreich – zu stellen. Die Betonung der Eigenverantwortung und das Entwickeln einer eigenen Stimme sind in Amsterdam sehr präsent.

Fabian Harb, Du arbeitest im Moment bei Laurenz Brunner – wie hat sich das ergeben und wie sieht Deine Arbeit dort aus? An welchen Projekten arbeitest Du mit?

fh: Ich habe während meines Studiums in der Kunsthalle in Sankt Gallen gearbeitet, für die Laurenz Brunner und Cornel Windlin[6] die Grafik bestellen. Als ich nach dem Studium zu meiner Freundin nach Amsterdam zog, war Laurenz meine erste Anlaufstelle. Der Einblick in seine Arbeit war ein Glücksfall und ich konnte nach einem Praktikum weiterhin an verschiedensten Projekten mitarbeiten.

In der Ausstellung »Call for Type« stellt ihr euren neuen Layer-Font Grow vor. Was war die Inspiration zu der Schrift? Könnt ihr kurz etwas zum Entstehungsprozess und zum Charakter der Schrift erzählen?

jb+fh: Grow ist unsere Reaktion auf ein Letraset-Schriftmuster, dessen räumliche Wirkung und Systematik uns fasziniert hat. Von dieser Grundlage ausgehend haben wir die Formen weiterentwickelt und das Character-Set auf 448 Buchstaben erweitert. Eine wichtige Komponente ist die Isolierung der einzelnen Outlines, die in der späteren Anwendung frei miteinander kombiniert werden können. Mit der Hilfe von Gustavo Ferreira,[7] einem Type-Engineer, der in der Produktion von Layer-Schriften sehr erfahren ist, hat Grow nun 63 Schnitte bekommen – jede erdenkliche Kombination der sechs Grundformen ist dabei.

Die Grow erscheint in der kürzlich von euch gegründeten Type Foundry Dinamo. Was war die Motivation, eine eigene Foundry zu gründen und wie stellt ihr euch die Zukunft von Dinamo vor?

Es ist durchaus eine fragwürdige Position geworden, noch eine weitere Foundry zu gründen und die typische Handvoll Schriften zu veröffentlichen. Andererseits haben sich Entwürfe in unseren Schubladen angesammelt, die bereits in einigen Projekten ihre Qualität beweisen und über verschiedene Anwendungen hinweg reifen konnten. Nachdem eigene Schriften den Dienst in ihren Projekten erfüllt haben, zirkulieren sie oftmals in unserem Bekanntenkreis. Besteht größere Nachfrage, ist es uns umso wichtiger, dass unsere Schriften auch aktuellste technische Anforderungen erfüllen.

Dass wir Grow unabhängig und mit keiner bereits bestehenden Foundry publizieren, ist vor allem unserer eigenen Perspektive geschuldet: Hätten wir einen geeigneten Platz gesehen, müsste es Dinamo nicht unbedingt geben.

Arbeitet ihr eher konzeptionell oder aus dem Prozess heraus?

Wir interessieren uns für spannenden Inhalt und einen gerechten Umgang damit. Maßgeschneiderte Schriften, die konkret auf den Projektkontext Bezug nehmen, können effektive Werkzeuge sein. Auf der anderen Seite ist es ebenso ein Anspruch, allgemeiner gültige Schriften zu entwerfen. In beiden Fällen ist für uns die Absicht zentral und die Form eine wichtige Folge davon.

Ist der Computer eher der Freund oder der Feind des Schriftgestalters? Welchen Einfluss hat die Wahl des Werkzeugs auf eure Arbeit?

Wenn wir über Produktion sprechen, ist der Computer ein hilfreiches Werkzeug, das

6 Cornel Windlin (*1964) besuchte die Grafikfachklasse an der Schule für Gestaltung in Luzern und arbeitete ab 1988 in London für Neville Brody. 1990 wechselte er als verantwortlicher Designer und Art Editor zum Magazin »The Face« und eröffnete ein Jahr später sein eigenes Studio. 1993 zog er nach Zürich, wo er unter dem Label »Lineto« eigene Schriftentwürfe veröffentlichte.

7 Gustavo Ferreira (*1977) lebt und arbeitet in Amsterdam. Grafik- und Produkt-Design-Studium an der Esdi in Rio de Janeiro, HfG Schwäbisch Gmünd und KABK Den Haag. Seit 2009 entwickelt er mit dem Design-Büro Hipertipo u. a. Software für die Gestaltung und Produktion von Schriften.

mühsame Arbeitsschritte beschleunigt und helfen kann, die Übersicht über komplexe Vorgänge zu bewahren. Geht es um Entwurf und Gestaltung, kann seine programmatische Funktionsweise aber den Blick auf interessantere Resultate versperren. Ohne Zweifel schafft der Computer auf dem Gebiet der Schriftgestaltung eine Bandbreite an neuen Möglichkeiten – andererseits leidet die Kopfarbeit schnell unter der Einfachheit des schnellen Klicks.

Was ist eure Haltung zum Thema Design-Trends? Wie beeinflussen sich zeitgemäße Schriftgestaltung und zeitgemäßes Grafik-Design?

Wie in jeder kreativen Disziplin gibt es auch in Grafik und Schriftgestaltung Tendenzen, die in gewissen Zeiten breiter spürbar sind. Man muss immer gut entscheiden, wie man sich selbst dazu verhält.

Im Kapitel »Call for Type« der vorliegenden Publikation werden auch ausgewählte Schriftentwürfe von 50 Newcomern vorgestellt. Was habt ihr jungen Schriftgestaltern mit auf den Weg zu geben?

Das Glück ist mit den Tüchtigen.

Das Interview mit Johannes Breyer
und Fabian Harb führten
Yvonne Kümmel und Jens Giesel.

Johannes Breyer[1] was born 1987 in Erlangen, Germany. After studying at the ZHdK[2] in Switzerland and an internship with the Zurich-based studio »Norm«,[3] he graduated 2013 from the Gerrit Rietveld Academie in Amsterdam.

Fabian Harb[1] was born 1988 in St. Gallen, Switzerland. After studying at the Basel School of Design, he has been working with Studio Laurenz Brunner[4] as well as persuing his own projects. Johannes Breyer and Fabian Harb jointly established the type foundry »Dinamo«.[1]

How did you get to know each other, and what projects have you completed together? How does your collaboration work?

jb+fh: We met in Switzerland through our mutual friend Larissa.[5] This soon resulted in an exchange between Basel and Zurich, which lasted throughout our study time. After graduating, Fabian moved to Amsterdam for an internship and Johannes continued his studies there. Living in the same place again, we started to work on projects and typefaces together. Today, we are both living and working in Berlin.

Do you see yourselves as type designers, or rather as graphic designers who create their own tools?

We see graphic design as a dialogue between all involved entities. This has to do with language, language is connected with type, and type with form. For us, it makes sense to shape even the smallest parts, bearing in mind the project's purposes. We start most of our commissions with the design of typefaces, but to answer your question: We are both trained as classic graphic designers – it would be rather bold to call ourselves primarily type designers.

Johannes, you are currently finishing your studies at the Gerrit Rietveld Academie in Amsterdam. What is it that fascinates you about the Rietveld, and why did you choose to study there? How do mentality and design at the Rietveld differ from Swiss or German universities and agencies?

jb: I benefit the most from the vivid exchange among teachers and students, who all come to Amsterdam from various countries for a short, but intensive period. This wide variety of interests and perspectives gives rise to many questions in an environment that provides enough room to address all possible kinds of ideas – and with varying success. A lot of emphasis is given to individual responsibility and the development of one's own position.

Fabian, you are at the moment working for the studio of Laurenz Brunner – how did this come about and what kind of work do you do there? What sort of projects are you working on?

fh: During my studies, I have been working with Kunsthalle in Sankt Gallen for which Laurenz Brunner and Cornel Windlin[6] are doing the graphic design. After graduation, I moved to Amsterdam and Laurenz' studio was the first place I went to. I gained insight into his practice and kept on working with him.

In the exhibition »Call for Type« you presented Grow, your new layer font. Where did your inspiration for this typeface come from? Can you tell us something about the design process and describe its characteristics?

jb+fh: Grow is our reaction to an old Letraset specimen whose layered effect and systematic construction seemed interesting to us. Based on that material, we developed the forms and completed the character set to a total of 448 glyphs. With the assistance of type engineer Gustavo Ferreira,[7] Grow grew to a system of 63 cuts – comprising all possible combinations of the six basic shapes.

Grow was published with Dinamo, the foundry you recently established. What was the driving force behind starting your own type foundry and what are your future plans with Dinamo?

It is indeed a questionable move to start yet another type foundry in order to publish the usual handful of fonts. On the other hand, typefaces that have already proven their qualities and have matured in the course of several applications keep piling up in our drawers. After having served their purposes at specific projects, these typefaces often start to circulate. If there is a greater demand for them however, it is all the more important that they

1 johannesbreyer.com
fabianharb.ch
dinamo.us

2 The Zurich University of the Arts resulted from the fusion in 2007 of the Academy of Design and Art (HGKZ) and the Academy for Music and Drama (HMT). The Zurich Design Museum is linked to the ZHdK (→ S. 28).

3 Norm was launched in 1999 by Dimitri Bruni and Manuel Krebs. Among their clients are Omega, Swatch and the Louvre. Manuel Krebs and Dimitri Bruni, who are known for their systematic approach in typography, have published some very successful typefaces via the type foundry Lineto (e.g. LL Simple, LL Replica.).

4 After studying in London (Central Saint Martins) and Amsterdam (Gerrit Rietveld Academie) Laurenz Brunner (1980) now lives and works in Amsterdam, Berlin and Zurich. Lineto published his LL Akkurat, his LL Circular came out in 2013.

5 Larissa Kasper studied Visual Communication at the ZHdK. Together with Rosario Florio she established the Design Studio Kasper-Florio in St. Gallen.

6 Cornel Windlin (*1964) studied Graphic Design at the Design School Luzern, Switzerland, and in 1988 started work for Neville Brody. In 1990 he became Art and Design Editor of the Magazine »The Face«, and a year later he started his own studio. In 1993 he moved to Zürich, where he publishes his own typefaces under the label »Lineto«.

7 Gustavo Ferreira (*1977) lives and works in Amsterdam. He studied graphic and product design at the Esdi in Rio de Janeiro, at the HfG in Schwäbisch Gmünd and KABK in The Hague. Since 2009 he has been developing software for type design and production in conjunction with the design studio Hipertipo.

meet with the latest technical standards. The reason for publishing Dinamo through none of the existing foundries has to do with our own perspective: If we had spotted a suitable platform for Grow to fit in, our own foundry would not have been necessary.

Is your work conceptual or is it more process orientated?

We are interested in exciting subjects and their adequate treatment. Bespoke typefaces reflecting the context of a project can be effective tools. On the other hand, it is equally challenging to create a universal typeface suitable for wider use. In both cases the intention is most important to us, and the form is a logical consequence.

Is the computer a friend or an enemy of the type designer? What impact has the tool on the outcome of your work?

When talking about production, the computer is a helpful tool which speeds up laborious production steps and helps to keep track of complex operations. Concerning the design work, its programmatic nature can also blank out interesting paths and results. The computer offers a whole spectrum of new possibilities; brain work and a healthy skepticism, however, can suffer from the »quick click«.

What do you think of design trends? How do contemporary type design and graphic design influence each other?

As in any creative discipline, there are definite tendencies in graphic and type design. You just have to decide how you want to position yourself.

The chapter »Call for Type« in this book presents a selection of type designs by 50 newcomers. What advice would you give to these young type designers?

Fortune favors the bold.

Johannes Breyer and Fabian Harb were interviewed by Yvonne Kümmel and Jens Giesel.

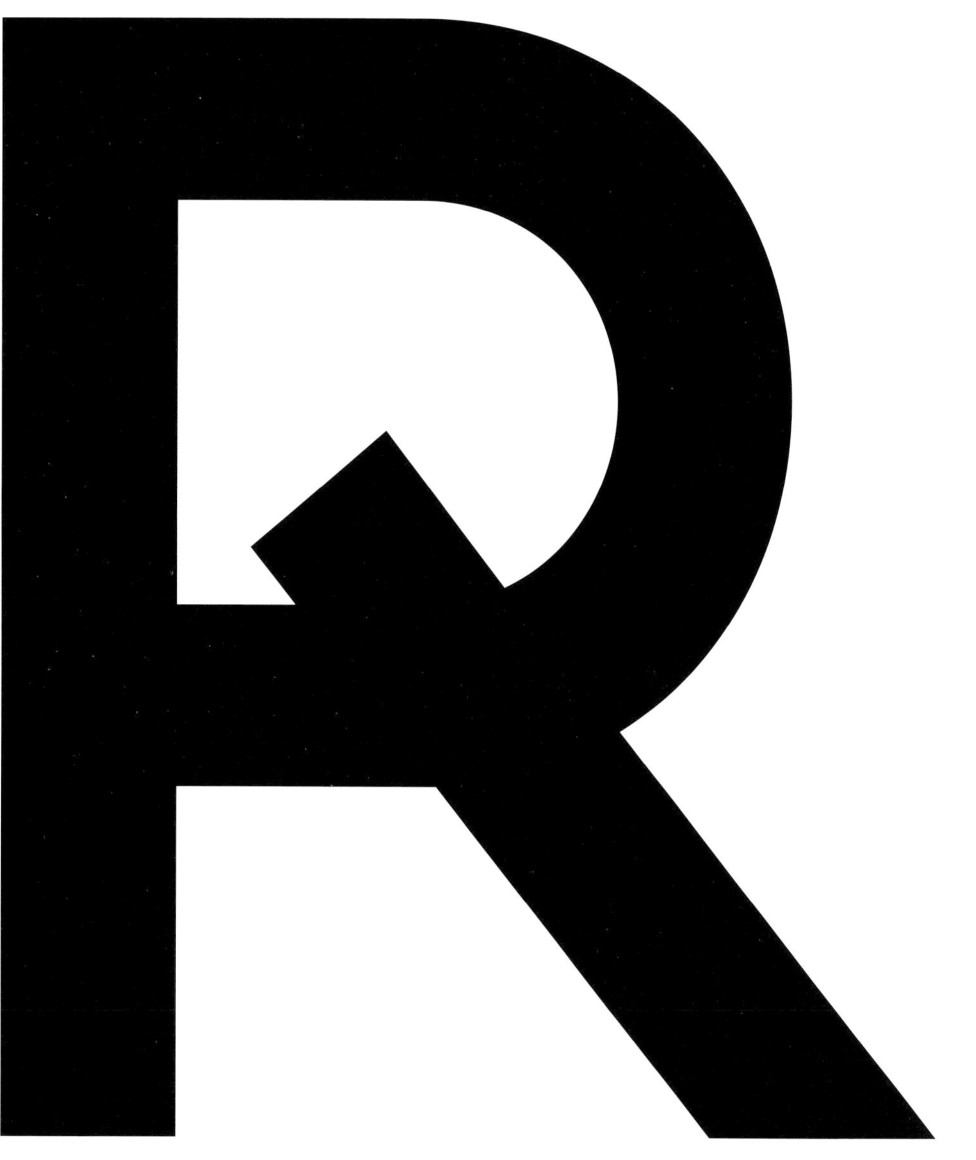

ABCDEFGHIJKLMNO
PQRSTUVWXYZ

abcdefghijklmnopqrstuvwxyz

0123456789

GT Haptik Medium

*ABCDEFGHIJKLMNO
PQRSTUVWXYZ*

abcdefghijklmnopqrstuvwxyz

0123456789

GT Haptik Medium Rotelic

Wenn es aber um das Leben gehe, sei das Tasten der primäre Sinn, da nichtstationäre Organismen auf den Tastsinn angewiesen sind.

GT Haptik Regular – 22pt

DIE RESTLICHEN SINNE DIENTEN ABER DEM GUTEN LEBEN, DA SIE

GT Haptik Medium Rotalic – 28pt

Orientierung ermöglichen

GT Haptik Medium – 64pt

Schrift. Typeface. GT Haptik
Gestalter. Designer. Reto Moser & Tobias Rechsteiner
Label. Foundry. grillitype.com
Jahr. Year. 2010

Die Schrift Haptik ist im Rahmen der Bachelor-Thesis »Optisch / Haptisch« an der Hochschule der Künste Bern entstanden. Sie bildete die Basis für das Kommunikationskonzept einer fiktiven Ausstellung, welche sich mit dem Tastsinn befasste. Ausgangsidee war es, eine möglichst tastfreundliche Schrift zu entwerfen. Dazu wurden die Versalien und Zahlen auf ihre einfachste Form heruntergebrochen und auf die erkennbare Funktionalität überprüft. Bei Tastversuchen wurden verschiedene Versionen und Varianten erprobt. Denn nicht nur die Buchstabenform, sondern auch die Strichstärke ist ein wichtiger Faktor beim Ertasten. Optische Korrekturen hingegen werden nicht wahrgenommen und schienen dadurch zunächst überflüssig. Um die Schrift jedoch auch in kleinen Punktgrößen einsetzen zu können, waren Strichstärkenkontrast und optische Korrekturen unumgänglich.

The type Haptik was developed within the framework of the Bachelor thesis »Optisch / Haptisch« at the University of the Arts in Bern / Switzerland. It represents the base for the communications concept of a fictitious exhibition dealing with tactile experience. The original idea was to create a type which was as tactile as possible. To this end, capital letters and numbers were broken down to their simplest form and tested for recognisable functionality. The different versions and variants were then tried out in haptic tests. It was found that not only the form, but also its thickness is an important factor when feeling a letter. Optic corrections, on the other hand, were not registered and therefore, at first, seemed superfluous. However, in order to use a smaller point type, contrasts in line thickness and optic corrections were indispensable.

Nach einer Polygrafen-Lehre studierte *Reto Moser*[1] (*1983) Visuelle Kommunikation an der Hochschule der Künste Bern. Er absolvierte Praktika bei »Folch Studio«[2] in Barcelona und »123buero«[3] in Berlin. Seit 2009 ist Moser als selbstständiger Gestalter in den Bereichen Editorial- und Schrift-Design tätig und arbeitet als Gestalter an der Hochschule der Künste Bern. Nebenbei unterstützt er seine Freunde bei der Type Foundry »Grilli Type«.[4]

***Tobias Rechsteiner*[1] (*1984) studierte bis 2009 Visuelle Kommunikation an der Hochschule der Künste Bern, absolvierte Praktika bei »Basedesign«[5] und »Cleo Charuet«[6] in Paris sowie bei »Gavillet & Rust«[7] in Genf. Mit dem Bachelor-Abschluss begann er im Bereich Schrift- und Editorial-Design zu arbeiten und war als Designer für die Berner Hochschule der Künste tätig. Er arbeitet für »Eclat«[8] und betreibt mit Freunden »Raum No«, ein freies Ausstellungsprojekt für zeitgenössische Kunst.**

Wie habt ihr euch kennengelernt?

tr: Das war schon vor dem Studium, ich war mit Retos Freundin auf dem Gymnasium. So haben wir uns schon ein halbes Jahr vor der Aufnahmeprüfung kennengelernt.

Hattet ihr schon im Studium euren Schwerpunkt auf Schriftgestaltung gelegt und woher kommt das Interesse für diesen Bereich?

tr: Bei mir kommt das Interesse definitiv vom Sprayen. Ich habe in meiner Gymnasialzeit mit der Spraydose … *(Pause)* Dinge gemacht *(lacht)*. Und irgendwann habe ich bemerkt, dass die Schrift das ist, was mich interessiert. Nachdem wir bereits im ersten Studienjahr mit Schrift zu tun hatten, sollten wir im zweiten Jahr im Fach Typografie eine Matrixfont zeichnen – und dazu hatten wir beide keine Lust, weil es uns banal erschien, irgendwelche Formen wild durcheinander zu stecken. So haben wir eigentlich schon beide unabhängig voneinander mit einer proportionalen Schrift experimentiert. Wir begannen uns für das Programm FontLab zu interessieren und haben uns selbst in die

1 Grotesk.cc

2 Folch Studio in Barcelona wurde 2004 von Albert Folch ins Leben gerufen.

3 2002 gründete Timo Gaessner das Berliner Design-Studio 123buero, das mit dem Schwerpunkt Typografie u.a. Ausstellungsdesign, Webseiten und maßgeschneiderte Schriften für Unternehmen entwickelt (→ S. 150).

4 Grilli Type: Schweizer Type Foundry, die 2009 von Noël Leu und Thierry Blancpain gegründet wurde.

5 Das Gestaltungsbüro Basedesign wurde 1997 gegründet. Es hat Büros in Genf, Brüssel, New York und Santiago.

6 Cleo Charuet: Pariser Grafik-Designerin.

7 2001 von Gilles Gavillet und David Rust ins Leben gerufenes Büro für Grafik-Design in Genf.

8 Die 1988 von Daniel Zentner gegründete Firma Eclat ist eine Schweizer Agentur für Markenberatung und Branding. Zu ihren Kunden gehören SBB und Swisscom (→ S. 22).

Materie eingearbeitet. Wir waren die einzigen in unserem Jahrgang, die sich so intensiv damit befasst haben und die Dozenten konnten uns hierbei kaum unterstützen.

rm: Da ich vor dem Studium eine Lehre als Polygraf absolviert habe, war ich vom typografischen Arbeiten geprägt und Schrift interessierte mich grundsätzlich – aber in der Lehre wurden uns auch die Grundlagen »eingetrichtert«. Als wir dann im Studium waren und freier an die Sache herangehen konnten, hat es mir extrem viel Spaß bereitet – und das macht es noch heute. Es ist einfach eine Arbeit, die mir sehr gut liegt.

Gibt es Projekte, bei denen ihr eure Schriften lieber nicht gesehen hättet?

(Beide lachen)
tr: Bis jetzt noch nicht. Wir kennen auch gar nicht so viele Anwendungen. Aber bis jetzt waren wir eher positiv überrascht. Wenn man die Schrift selbst gestaltet, schränkt man sich in der Anwendung ein – weil man die Schrift eben so gut kennt, und dann gibt uns das eher neuen Input, wenn andere sie benutzen.

Denkt ihr, dass die Qualität von Schriftentwürfen durch die heutige vereinfachte Zugänglichkeit zum Schriftentwurf leidet oder eher profitiert?

rm: Beides. Obwohl ich eher zum Ersteren tendieren würde. Wenn ich mir anschaue, was ich seit Entwerfen der Haptik dazu gelernt habe, muss ich sagen, dass es die Software alleine nicht ausmacht. Es ist extrem wichtig, dass man sich lange und intensiv damit beschäftigt. Aber natürlich: Je eher man die Software versteht, desto schneller kommt man zum Kern des eigentlichen Gestaltens.

tr: Es ist wahrscheinlich so wie mit Photoshop. Es gibt einfach mehr Schund, weil es Photoshop gibt, aber zusätzlich hat natürlich mit den Möglichkeiten, die das Programm bietet, auch die Qualität der Bilder massiv zugenommen. Es ist einfach so, dass viel mehr Leute Schriften machen. Aber ich würde nicht sagen, dass es prozentual mehr Schund gibt. Hm… oder vielleicht doch.

rm: Es ist heute einfach so, dass man eher noch schnell eine Schrift macht. Dann kommt es natürlich darauf an, wie gut man das beherrscht. Wenn ich mir meine ersten beiden Schriftprojekte anschaue, dann muss ich mich heute am Kopf kratzen. Aber das ist vermutlich immer so.

Gab es für die Haptik einen bestimmten Anlass?

tr: Die Haptik war ein Teil unserer Bachelor-Thesis. Dafür haben wir uns mit der Gestaltung für Sehbehinderte beschäftigt. Wir haben also zuerst recherchiert, wie die Kommunikation mit Blinden stattfinden kann und haben uns dann entschieden, die Kommunikation für eine fiktive Ausstellung zu gestalten. Eine Ausstellung, die sich mit tastbarer Kommunikation beschäftigt. Dazu haben wir dann diese Schrift entwickelt.

Was macht die Schrift aus und wie funktioniert sie?

rm: Das Besondere an der Schrift ist, dass die Versalien keine optischen Korrekturen aufweisen, da dies für das Ertasten keine Rolle spielt. Die Kleinbuchstaben kamen im Verlauf des Projekts dazu und sind einfach ergänzend, um einen lesbaren Fließtext setzen zu können. Diese wiederum weisen alle nötigen optischen Korrekturen auf. Es hat sich auch herausgestellt, dass die Kleinbuchstaben zu schwierig zu ertasten sind. Dadurch, dass die Großbuchstaben fürs Ertasten optimiert sind, entstehen diese seltsamen Formen. Wie zum Beispiel beim großen G. Da haben wir gemerkt, dass es keine ganze Rundung braucht, um es tastbar zu machen. Das haben wir lange mit Hilfe eines speziellen Papiers ausprobiert und waren dann bald soweit, dass wir wussten, was funktioniert und was nicht.

Wie lange habt ihr an der Schrift gearbeitet?

tr: Die Schrift ist in nur zweieinhalb Wochen entstanden. Natürlich haben wir sie, als wir sie später zum Verkauf anbieten wollten, noch einmal überarbeitet, das Spacing und Kerning optimiert, den Zeichensatz erweitert und so nochmal sehr viel Zeit investiert, aber die Schrift an sich war nach den zweieinhalb Wochen schon fertig.

»Es ist einfach so, dass viel mehr Leute Schriften machen. Aber ich würde nicht sagen, dass es prozentual mehr Schund gibt. Hm … oder vielleicht doch.«

Tobias Rechsteiner
→ S. 99

rm: Ich habe vor allem gezeichnet und wir haben sie uns dann immer wieder zusammen angeschaut und das hat circa zweieinhalb Wochen gedauert.

Habt ihr Hilfe beim Ausarbeiten eurer Entwürfe?

rm: Wir machen alles selbst – ich geb' meine Sachen auch nicht so gerne aus der Hand.

tr: Also, wenn ich genug Geld verdienen würde, um jemanden zu bezahlen, würde ich die Arbeit, die kommt, wenn der ganze Zeichensatz gezeichnet ist, wie etwa die Akzente, das Kerning, Spacing und den ganzen anstrengenden Teil, liebend gerne abgeben *(lacht)*. Das geht aber im Moment nicht.

Ihr arbeitet in verschiedenen Städten. Wie sieht dann die Arbeit an gemeinsamen Projekten aus?

tr: Ich arbeite und lebe in Zürich. Deshalb ist das nicht so einfach. Aber da wir einfach sehr gute Freunde sind, sehen wir uns regelmäßig. Wir arbeiten momentan eher an eigenen Projekten und geben uns dann gegenseitig Feedback übers Internet, aber wir setzen uns nicht zusammen hin und erarbeiten etwas. Wenn wir uns treffen, dann weil wir privat Zeit miteinander verbringen. Lustigerweise unterhalten wir uns dann auch nicht mehr viel über Gestaltung – das haben wir früher immer gemacht. Es gab mal eine idealistische Zeit *(beide lachen)*.

Ist für euch die Schweiz als Standort und Inspiration wichtig?

tr: Ich würde sagen, dass das ortsunabhängige Arbeiten kein Problem ist. Der Einfluss der Schweiz auf uns ist aber sicher sehr stark. Man wächst mit der Beschilderung der schweizerischen Bundesbahn und dem Schweizer Rationalismus auf – das ist dann einfach drin. Ich bin im Moment häufig in Berlin, weil meine Freundin da lebt. Deutsches Design ist definitiv anders. Es wäre für mich schon interessant dort zu arbeiten.

rm: Ich bin hier, weil meine Leute hier sind – wo ich arbeite, ist eigentlich irrelevant.

tr: Die Szene in der Schweiz ist sehr klein und irgendwann lässt dieser Einfluss in dem Sinne nach, dass man sich dadurch inspiriert fühlt. Dennoch wäre es sicher inspirierend, irgendwo anders hinzuziehen und neue Einflüsse zu erleben.

Wie kam es dazu, dass ihr die Haptik bei Grilli Type zum Verkauf anbietet?

rm: Thierry Blancpain[9] und Noël Leu[10] haben ein Jahr unter uns studiert, und da an der Hochschule im Bereich Schriftgestaltung nicht unseren Vorstellungen und Interessen entsprechend unterrichtet wurde, haben wir uns zusammen mit den beiden selbst organisiert. Wir haben eine Art Selbsthilfegruppe gegründet *(lachen)*. Nach dem Studium haben die beiden Grilli Type ins Leben gerufen und wir waren die ersten, die eine Schrift beigesteuert haben.

In den letzten Jahren sind immer mehr Type Foundries entstanden. Glaubt ihr, dass das eine nachhaltige Entwicklung oder eher ein Trend ist?

tr: Ich würde behaupten, dass von den Foundries, die in den letzten zwei Jahren gegründet wurden, in zwanzig Jahren noch zwanzig bis dreißig Prozent übrig sind.

rm: Auf der anderen Seite liegen die Schriften ja einfach rum, d.h. wenn man diese zum Verkauf anbietet, muss man sich eigentlich nicht mehr darum kümmern. Ich denke, es ist wie in der Musik, jeder möchte sein eigenes Label haben, und wenn man selber Schriften macht, möchte man auch selbst darüber entscheiden, wie und wo diese vertrieben werden. Ich kann mir vorstellen, dass das irgendwann zurückgeht, zumal Typo im Moment auch sehr trendy ist.

9 Thierry Blancpain (*1985), Schweizer Brand- und Interaction-Designer, der 2009 zusammen mit Noël Leu die Foundry Grilli Type initiierte.

10 Noël Leu (siehe 9)

tr: Ich denke, dass die großen Foundries, wie Monotype oder Fontshop, gerade merken, dass ihr Modell, bei dem teilweise nur zwanzig Prozent der Einnahmen an den Designer gehen, nicht mehr lange funktioniert und das wäre auch zu hoffen. Bei den kleineren Foundries gibt es oft ein sehr viel faireres Entlohungsverfahren und das ist mit Sicherheit eine Reaktion darauf.

Habt ihr Vorbilder?

tr: So wie Justin Biber *(lacht)*? Reto ist mein Vorbild *(lacht)*. Was Agenturen angeht, ist mein Vorbild vielleicht Wolff Olins.[11] Die machen Branding auf einem sehr hohen Niveau, zwar sehr kommerziell, aber extrem gut. Ich denke, die Vorbildzeit ist auch ein bisschen vorbei, die hatte man während des Studiums, wo man sich auch noch selbst findet. Jetzt bewundere ich eher Leute, die für mich viel konkreter sind und nicht mehr so unerreicht. Zum Beispiel Freunde von uns: Dimitri und Noah, die eine eigene Agentur namens B & R Grafikdesign[12] gegründet haben. Das beeindruckt mich.

Was war euer anstrengendstes Projekt bisher?

(Lachen)
tr: Unser erstes Projekt nach dem Studium. Da mussten wir über eine Marketing-Agentur das Erscheinungsbild für eine Sanitärinstallationsfirma machen. Das Katastrophale war, dass wir die Entwürfe machen sollten und die vom Marketing dann damit zum Kunden gegangen sind. Wir hatten von Anfang an gesagt, dass wir beim Treffen mit dem Kunden dabei sein wollen – aber das wollten die nicht. Und dann kamen die zurück und meinten, wir sollten das neu machen. Das hat sich dann nochmal wiederholt und beim dritten Mal haben wir ihnen 100 Logoentwürfe mitgegeben *(lachen)*. Und das Beste war, dass dem Kunden nichts gefallen hat. Dann hat die Agentur den Designer gewechselt und die haben ein Logo gewählt, das katastrophal ist und in unseren Entwürfen bereits drin war! Also ein Tipp: Macht so etwas nie!

Das Interview mit Reto Moser und Tobias Rechsteiner führten Lynn Blees und Luzia Hein.

11 Die Brandingagentur Wolff Olins wurde 1965 von Michael Wolff und Wally Olins in London gegründet. Zu ihren Kunden gehören Unilever, Mercedes-Benz und Windows.

12 B & R Grafikdesign ist ein Gestaltungsbüro in Bern von Dimitri Reist und Noah Bonsma.

After an apprenticeship as a typographer and lithographer, *Reto Moser*[1] (*1983) studied Visual Communication at the Bern University of the Arts. He had an internship with »Folch Studio«[2] in Barcelona and »123buero«[3] in Berlin. Since 2009 he has been working as a freelance designer for editorial and type design, he also works as designer at the Bern University of the Arts and supports his friends at the type foundry »Grilli Type«.[4]

Until 2009 *Tobias Rechsteiner*[1] (*1984) studied Visual Communication at the Hochschule der Künste, Bern, then had internships with »Basedesign«[5] and »Cleo Charuet«,[6] Paris, as well as with »Gavillet & Rust«[7] in Geneva. After his B.A. he started work in Type and Editorial Design and worked for the Bern University of the Arts. He works for »Eclat«[8] and together with friends runs »Raum No«, a free exhibition project for contemporary art.

How did you get to know each other?

tr: We met before our studies. I was in the same secondary school as Reto's girlfriend, and so we got to know each other six months before our qualifying examinations.

Did you already focus on type design during your studies? How did you get interested in this field?

tr: It was definitely graffiti art that sparked my interest in type design. When I was a secondary school student, I used a spray can … *(pause)* to create things. *(He laughs)*. Eventually I realized that it was type that really interested me. After having dealt with type design already in our first academic year, we were told in our typography course of the second year to draw a matrix font. The two of us didn't like this idea because we thought it was banal to put several forms together in a haphazard way. Therefore we both experimented with a proportional font independently. We started being interested in FontLab and learnt how to use it on our own. We were the only students of that academic year who occupied themselves so intensely with FontLab, and our university teachers couldn't support us a lot.

rm: Since I had been trained as a typographer and lithographer before my studies, I was influenced by typographic work and type was something that fascinated me. However, the basics had been »spoon-fed« to us during our training. When we began our studies and could approach things more independently, I really enjoyed it – and I still enjoy it today. It is a kind of work that really suits me.

Were there any projects where you disliked the use of your fonts?

(They start laughing).
tr: No, not yet. Actually we don't know very many applications of them. So far we have been rather pleasantly surprised. When you create a font on your own, you personally reduce its scope of application because you just know your font so well. So we get new creative input, if we see how our font has been used by other people.

Do you think that the quality of type design suffers from today's easier access to it or that the quality rather benefits from that fact?

rm: Both. However, I tend towards the first option. When I look at the new things I have learnt since designing Haptik, I have to admit that it's not the software alone which is essential. It's vitally important to immerse yourself into type design intensely and for a long time. However, the sooner you understand the software, the quicker you get to the core of real design.

tr: Probably it's similar to Photoshop. Actually there's considerably more trash because of Photoshop. At the same time, however, this programme offers more possibilities, so the quality of images has significantly increased. There are more people who create fonts – it's as simple as that. However, this doesn't mean that there's proportionally more rubbish. Um … or maybe it does after all.

rm: You're much more likely to create another font on the fly – that's just how it is today. Of course it then depends on how good you are at it. Today, my first two typeface projects make me scratch my head. But that's probably how things are.

Was there any special reason for the creation of Haptik?

tr: Haptik was part of our bachelor thesis. For this, we looked at design for the visually impaired people. At first, we researched how communicating for the blind takes place. Then we decided to create a communication for a fictitious exhibition, that deals with tactile communication. Therefore we developed this font.

What are the characteristics of this font, and how does it work?

rm: The special thing about this font is that the majuscules don't have any optical corrections because they are not important for the sense of touch. During the project we added minuscules to the font, to compose legible running text. The minuscules, however, show all necessary optical

1 Grotesk.cc

2 Folch Studio in Bacelona was established in 2004 by Albert Folch.

3 In 2002 Timo Gaessner established the Berlin design studio 123buero with the emphasis on typography, which develops exhibition design, websites and tailor-made typefaces for companies (→ S. 155).

4 Grilli Type: Swiss type foundry, established in 2009 by Noël Leu and Thierry Blancpain.

5 The design studio Basedesign was founded in 1997. It has branches in Geneva, Brussels, New York and Santiago.

6 Cleo Charuet: Paris graphic designer.

7 Gavillet & Rust is a studio for graphic design, established in 2001 by Gilles Gavillet and David Rust.

8 Eclat, established in 1988 by Daniel Zentner, is a Swiss branding agency. Among their clients are SBB and Swisscom (→ S. 28).

corrections. It also became apparent that it was too difficult to read the minuscules blindfolded. Since the majuscules are user-optimized, they show somewhat strange letter forms. For example, if you look at the capital letter G. We've noticed that this letter did not have to be completely round if you wanted to feel it. This is what we found out after long experiments with a special paper. Soon we knew the things which would work and those which wouldn't be possible.

How long did it take to develop this font?

tr: The font was created in only two and a half weeks. Of course it needed some reworking when we decided to sell it. Later we optimized the letter-spacing and kerning and expanded the character set. So we invested a lot of additional time. However, the design itself was finished after two and a half weeks.

rm: I mainly did the drawing and we looked it over together. This took us about two and a half weeks.

Does anyone help you with the elaboration of your designs?

rm: We do everything on our own – I'm also not particularly keen on leaving my things to others.

tr: Well, if I earned enough money to pay somebody, I would really like to leave the work which remains after the drawing of the whole character set to others, such as diacritics, kerning, spacing, and the whole exhausting part. *(He laughs)*. But that's not possible at the moment.

The two of you work in different cities. How do you manage to work on mutual projects?

tr: I work and live in Zurich. This is why it's not that easy. But since we are very good friends, we meet regularly. At the moment we mostly work on individual projects and give each other feedback over the Internet. But we don't sit down together in order to work on something. When we meet, it's for private reasons. Oddly enough, we then don't talk a lot about design anymore – this is what we used to do some time ago. Those were the more idealistic times. *(Both laugh)*.

Is Switzerland important to you as a location and a source of inspiration?

tr: I think that location-independent working doesn't cause any problems. However, Switzerland has certainly had a big influence on us. One grows up with the signposting of the Swiss Railway and Swiss rationalism. – It's just something you have internalized. At the moment I'm frequently in Berlin because my girlfriend lives there. German design is definitely different. Actually it would be interesting to work there.

rm: I'm here because my family and friends are here – it doesn't actually matter to me where I work.

tr: The Swiss scene is very small, and sooner or later its influence decreases in the sense of finding inspiration in it. However, it would certainly be inspiring to move to another place and experience new influences.

How did it happen that you decided to sell Haptik via Grilli Type?

rm: Thierry Blancpain[9] and Noël Leu[10] studied at the same school in the class below us. Since the University didn't offer any training in the field of type design that could meet our expectations and interests, we organized something on our own. We established a kind of a self-help group *(they laugh)*. After their studies, Blancpain and Leu founded Grilli Type; and we were the first ones to contribute a font to this enterprise.

In recent years the number of type foundries has steadily increased. Do you think that this is an enduring development or just a trend?

tr: I suppose that in twenty years only twenty to thirty percent of the type foundries that have been established in the last two years will be remaining.

rm: On the other hand, the fonts are there. I mean, once you offer them for sale, you needn't look after them anymore. I think it's like in the music industry: everybody wants to have his own label. And if you create your own fonts, you also want to decide how and where they are sold. I can imagine that this development could decrease at some point in the future, especially since typography is very trendy at the moment.

tr: I think that big foundries like Monotype or Fontshop are beginning to realise, that their method of giving only 20 percent of the earnings to the designers won't be feasible for much longer; and I also hope that this will come to an end. Small foundries often have salary settlements which are much fairer; and that certainly is a reaction to the practices by the major foundries.

Do you have any role models?

tr: Somebody like Justin Biber? *(He laughs)*. Reto is my role model. *(He laughs)*. In terms of agencies, it's possibly Wolff Olins.[11] The work in branding they do is at a very high level. It's very commercial but well done. I think the times where we had role models are over. We had role models in our student days when we were still trying to find ourselves. Now I admire people who are much more concrete and more approachable like, for example, our friends Dimitri und Noah. They have their own agency called B & R Grafikdesign.[12] That impresses me.

What has been your most exhausting project so far?

(Both laugh).
tr: Our first project after our studies. A marketing agency wanted us to create a corporate design for a sanitary installations-company. It was a disastrous undertaking because we had to create the design, and then the marketing people showed it to the customer. We had told the marketing people right from the start that we wanted to attend their meeting with the customer. However, they didn't want to accept that. When they came back, they said that we should remake it. This happened again, and the third time we gave them 100 design concepts for logos *(they laugh)*. And to top it all off, the customer didn't like any of them. Then the agency entrusted another designer with the task, and they chose a logo which was dreadful and had already been among our concepts. So let me give you a piece of advice: Don't ever do anything like that!

Reto Moser and Tobias Rechsteiner were interviewed by Lynn Blees and Luzia Hein.

9 Thierry Blancpain (*1985), Swiss Brand and interaction designer, who founded Grilli Type in 2009, together with Noël Leu.

10 Noël Leu (see 9)

11 The branding agency Wolff Olins was founded in 1965 in London by Michael Wolff and Wally Olins. Among their clients are Unilever, Mercedes-Benz and Windows.

12 B & R Grafikdesign is a design studio in Bern run by Dimitri Reist and Noah Bonsma.

»Wir haben eine Art Selbsthilfe-Gruppe gegründet.«

GT HAPTIK

Reto Moser
→ S. 101

Happy Birthday!

Just Married!

Merci Beaucoup!

Congratulation!

Schrift. Typeface. Happypeppy
Gestalter. Designer. André Gröger & Susanne Kehrer
Label. Foundry. ilikebirds.de
Jahr. Year. 2010

Der Entwurf dieser Schrift wurde inspiriert von der Idee, eine Schrift für ein besonderes Glücksgefühl zu entwickeln. Konkretes Ausgangsmaterial für das Experiment war Geschenkband, das als Metapher für Überraschungen und Geschenke steht. Dabei war es den Gestaltern wichtig, die ursprünglichen Eigenschaften und Besonderheiten des Materials beizubehalten und in die fertige Schrift einzubeziehen. So entstand der eigenartig individuelle Schriftcharakter, der diese Schrift auszeichnet.

Experimentelle Headline-Schrift, 26 Zeichen (Minuskeln und Majuskeln), Nummern plus Sonderzeichen.

This type design was inspired by the idea to create a type for a feeling of happiness. The actual starting material for this experiment was gift ribbon, a symbol for presents and surprises. Here, it was important for the designers to try and preserve the original properties of this material and to include it in the final result. Thus, a very unique and distinctive typeface was born.

Experimental Headline font, 26 letters (minuscule and majuscule), numbers and additional characters.

Nach ihrer Ausbildung zur Mediengestalterin bei der Verlagsgruppe Rhein-Main studierte *Susanne Kehrer*[1] (*1979) Kommunikationsdesign an der FH Mainz und verbrachte ihr Auslandssemester an der Inholland University in Rotterdam. Sie war Wissenschaftliche Mitarbeiterin an der FH Mainz, bevor sie 2010 mit André Gröger das Designstudio »I LIKE BIRDS«[2] gründete. Für ihre Kunden entwerfen die beiden Gestalter Printprodukte, visuelle Identitäten und Konzepte. Ihre Illustrationen und Installationen wurden in Magazinen wie »Neon«, »Zeit Campus« oder »Kinki« publiziert.

Nach dem Studium an der Frankfurter Akademie für Kommunikation und Design (2001–2003) studierte *André Gröger*[1] (*1979) Kommunikationsdesign an der FH Mainz, absolvierte ein Auslandssemester an der Swinburne University in Melbourne (Australien) sowie ein Praktikum bei »HORT«[3] und arbeitete ein Jahr lang als freier Grafik-Designer. 2010 gründete er mit Susanne Kehrer das Designstudio »I LIKE BIRDS«.

Woher kennt ihr euch und wann habt ihr beschlossen zusammenzuarbeiten?

sk: Kennengelernt haben wir uns während unserer Diplomzeit, eigentlich in der letzten Woche, bevor wir unser Studium an der Fachhochschule Mainz beendet hatten. Danach hatte ich eine Assistentenstelle an der FH und André war zunächst als Freelancer tätig. Das ging ein Jahr. Danach beschlossen wir, uns zusammenzutun und ein Studio zu gründen. Anfangs waren wir erst einmal ein halbes Jahr in Mainz. Aber wir hatten schon immer im Blick, die Stadt zu wechseln und sind dann nach Hamburg gezogen. Das ist jetzt drei Jahre her. Seitdem sind wir hier.

War es eine bewusste Entscheidung, nach Hamburg zu gehen und nicht in eine andere Stadt? Findet ihr Hamburg als Standort wichtig?

ag: Ich war während meiner Studienzeit fünf Jahre in Mainz, Susanne ist von dort. Es war schon klar, dass wir nach dem Studium an einen anderen Ort wollten. Wir haben geschaut, ob wir nach Berlin gehen oder eben nach Hamburg. Dazu haben wir beide Städte bereist und sie miteinander verglichen. In Hamburg war es vom Gefühl her einfach schöner für uns. Ich kann auch nicht genau sagen, woran es konkret lag – man kommt in eine Stadt und fühlt sich wohl. Hamburg fanden wir entspannter und gemütlicher als Berlin.

1 ilikebirds.de

2 I LIKE BIRDS ist ein 2010 von Susanne Kehrer und André Gröger gegründetes Büro für Gestaltung in Hamburg.

3 Der Berliner HORT wurde 1994 von Eike König gegründet und ist mittlerweile ein multidisziplinäres Design-Studio, das international für große wie kleine Kunden arbeitet.

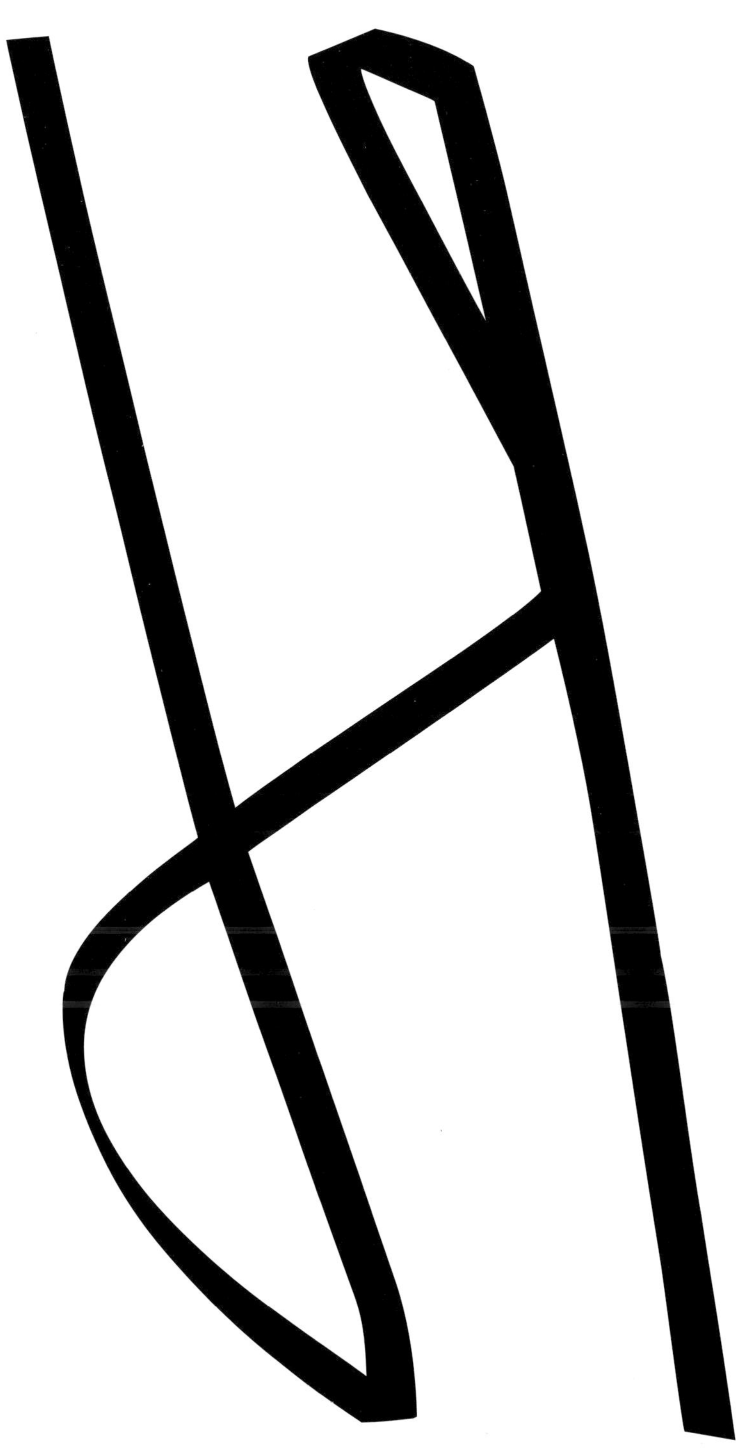

Im Vergleich zu anderen in der Ausstellung »Call for Type« vertretenen Font-Designern ist euer Umgang mit Schrift oftmals experimentell. Wo ordnet ihr euch als Schriftgestalter ein?

sk: Die Happypeppy ist ein experimenteller Font. Also nicht unbedingt dafür gedacht, Fließtexte zu setzen, sondern eher ein Headline-Font. Uns ging es um die Umsetzung. Wir haben die einzelnen Buchstaben aus Geschenkband geformt, dann abfotografiert, gescannt und digitalisiert. Dabei war uns der Prozess und die Digitalisierung des Handgemachten besonders wichtig.

ag: Es war ein Versuch. Ich hatte zuvor noch keine Schrift entworfen, aber großes Interesse daran. Die Umsetzung ist geprägt durch eine eher spielerische Herangehensweise. Uns war klar, dass wir das Ergebnis nicht als Leseschrift verwenden wollen. Wir sehen uns selbst nicht als Font-Designer. Schriftgestaltung ist für uns eher ein Bereich, in den wir immer wieder eintauchen.

Gab es einen bestimmten Anlass für die Happypeppy?

ag: Sie entstand als freies Projekt, als Auseinandersetzung mit dem Thema. Wir hatten eine Idee, einen Ansatz, den wir verfolgen wollten. Es hat sich dann zufällig ergeben, dass die Schrift für den Typodarium-Kalender ausgesucht und publiziert wurde. Das hat uns gefreut.

Habt ihr für die Happypeppy Anwendungen gesehen?

sk: Wir wissen, dass eine Bekleidungsfirma die Schrift für ihr Lookbook benutzen möchte und finden es natürlich schön, wenn auch andere Gefallen an der Schrift finden.

Also eher positive Erfahrungen?

ag: Ja, es hat sich niemand beschwert. Es ist ja immer gut, wenn Du eine Sache produzierst, mit der Du zufrieden bist und an der auch andere Leute Gefallen finden. Wir haben keine konkreten Anwendungsbeispiele bekommen, aber die Schrift wurde öfter gekauft. Es wäre ja schade, eine Schrift nur für uns zu entwickeln. Als wir fertig waren, dachten wir: Wir haben diese Schrift jetzt gemacht – warum sollten andere Leute nicht auch etwas davon haben?

Habt ihr bestimmte Inspirationsquellen für eure Arbeiten oder Rituale um Ideen zu finden?

sk: Inspirationen holen wir uns eher aus dem Kunstbereich. Weniger bei »zeitgemäßen« Quellen wie zum Beispiel Blogs im Internet. So etwas schauen wir grundsätzlich sehr selten an, im Gegensatz zu Büchern und speziellen Zeitungen. Oder wir gehen ins Museum. Der Alltag bietet uns ebenfalls viele Anregungen.

Arbeitet ihr lieber analog oder digital?

sk: Ich glaube, das eine geht nicht ohne das andere. Wir fangen meistens analog an, die Digitalisierung ist dann der nächste Schritt. Dieses Prinzip wenden wir bei den meisten Projekten an.

ag: Es ist für uns eine logische Folge, weil wir nicht den ganzen Tag am Computer sitzen können bzw. wollen. Deshalb schauen wir, wie wir ein bisschen per Hand arbeiten, um alles auf eine andere Ebene zu bringen. Das finden wir individueller, weil man nicht irgendetwas produziert, das in einem Computerprogramm vorgegeben ist. Man hat ja selbst am Rechner irgendwann bestimmte Einschränkungen beim Gestalten. Daher birgt die Mischung zwischen analog und digital mehr Möglichkeiten.

Ihr arbeitet ja auch viel für Magazine wie Neon, Kinki oder Zeit Campus. Wie kam es dazu?

ag: Im Magazin-Bereich ging es eigentlich mit der Neon los. Das kam über ein paar Ecken, weil man sich über bestimmte Kontexte schon kennt. Das war der Anfang …

sk: Na ja, Du kennst eben Leute, das ist in unserer Branche recht wichtig, um ein Netzwerk aufzubauen. Oder Du hast schon bestehende Kontakte, dadurch trägt sich das Ganze dann einfach weiter. Manchmal können wir das selbst nicht zurückverfolgen. Aber wir haben festgestellt, letztlich besteht dann immer eine Verbindung zu uns.

ag: Du landest in dem einen Magazin, das dann andere Leute sehen, die das weitertragen und daraufhin melden sich wieder Leute. Es ist nicht nur so, dass wir sagen, wir kennen den, der arbeitet dort, und den, der arbeitet da, und deswegen kam es zu den und den Aufträgen. Es ist immer eine Überraschung, wenn Leute sich melden und wir fragen natürlich auch nicht nach: Wie seid ihr auf uns gekommen? Das ist ein automatischer Ablauf.

Gibt es Projekte, die euch seit Langem reizen, oder schon lange in der Schublade liegen?

ag: Mhh… ich überlege gerade, was so in den Schubladen ist *(lacht)*.

sk: Eigentlich hatten wir immer das Glück, mit Kunden zusammenzuarbeiten und in diesem Zusammenhang alles zu verwirklichen. Wir hatten unglaublich viel Freiraum und konnten immer das machen, was wir wollten. Von daher haben wir jetzt keine geheime Schublade, in der≈noch Sachen liegen. Wir sind offen für alles. Es ist spannend zu wissen, dass sich jeden Tag jemand melden und uns mit interessanten Projekten überraschen könnte. Von daher schauen wir mal, was kommt.

ag: Spezielle Wunschprojekte haben wir momentan nicht, und meine Schubladen sind eigentlich auch alle aufgeräumt *(lacht)*.

Das Interview mit Susanne Kehrer
und André Gröger (I LIKE BIRDS)
führten Lynn Blees und Luzia Hein.

After training as a media designer with the publishing group Rhein-Main, *Susanne Kehrer*[1] (*1979) studied Communication Design at the University of Applied Sciences, Mainz. Her semester abroad she spent at the Inholland University in Rotterdam. She was a research assistant at the FH Mainz and in 2010 she established, together with André Gröger, the design studio »I LIKE BIRDS«,[2] where the two designers created print products, visual identities and concepts for their clients. Their illustrations and installations were publicized in magazines like »Neon«, »Zeit Campus« or »Kinki«.

After studying at the Frankfurt Academy for Communication and Design (2001-2003), *André Gröger*[1] (*1979) studied Communication Design at the University of Applied Sciences in Mainz, his semester abroad he spent at Swinburne University in Melbourne, Australia. After an internship with »HORT«[3] he spent a year working as a freelance graphic designer. In 2010 he established the design studio »I LIKE BIRDS« together with Susanne Kehrer, creating print products, visual identities and concepts for clients.

Where did you two meet and when did you decide to work together?

sk: We got to know each other during the time when we did our degrees, actually in the last week before we finished our studies at the Mainz University of Applied Sciences. Afterwards I worked as an assistant lecturer at the UAS whereas André began to work freelance. We did this for one year. Then we decided to join forces and establish a studio. Initially we were in Mainz where we stayed for six months. But we always wanted to go to another city. Finally we moved to Hamburg. This happened three years ago, and we have been here ever since.

Was it a conscious choice to go to Hamburg rather than any other city? Do you think that Hamburg is an important place?

ag: During my studies I spent five years in Mainz, and Susanne is from there. Actually it was quite clear that we wanted to go to another place after our studies. We wondered whether we should go to Berlin or Hamburg. So we visited both cities and compared them. We had a feeling that Hamburg was more appealing to us. I can't give you any concrete reason for that. – You just go to a city and it makes you feel at ease. We had the impression that Hamburg was cosier and more relaxed than Berlin.

Compared to the other type designers who are represented in the exhibition entitled »Call for Type«, your approach to typefaces is often experimental. Where do you see yourselves as type designers?

sk: Happypeppy is an experimental font. It isn't necessarily meant for the composition of running text but rather it is a headline font. It was its realization which mattered to us. We formed ribbons into single letters. Then we photographed, scanned, and digitized them. It was the whole process and the digitizing of something handmade that we found particularly important.

ag: It was an experiment. I had never created a font before but I was much interested in doing it. The creation of this font was considerably influenced by a rather playful approach. It was clear to us that we didn't want to use our result as a body copy font. We don't consider ourselves type designers. We rather see type design as a field we like to delve into from time to time.

Was there any special reason for the creation of Happypeppy?

ag: It was created as a free project based on our interest in this subject. We had an idea or a basic plan which we wanted to carry out. It just so happened that the font was chosen for the Typodarium calendar and published in it. We were happy about that.

Have you seen any examples where Happypeppy was used?

sk: We know that a clothing company wants to use the font for its look book. Of course we are glad to see that others like our font, too.

So you had rather positive experiences?

ag: Yes, nobody has complained so far. Actually it's always good to produce something that you are happy about and also pleases other people. We haven't got any concrete examples showing its use. However, the font was sold several times.

1 ilikebirds.de

2 I LIKE BIRDS is a design studio in Hamburg, founded in 2010 by Susanne Kehrer and André Gröger.

3 The HORT in Berlin was founded in 1994 by Eike König and has developed into a multidisciplined design studio working for a large range of international clients.

In fact, it would be a pity if we created a font only for ourselves. When it was finished, we thought: Now we have produced this font – why shouldn't other people benefit from it, too?

Have you any special sources of inspiration for your work or any rituals helping you to develop ideas?

sk: We are rather inspired by art but not so much by »contemporary« sources like blogs on the Internet. In fact, it's very rare that we read such things, in contrast to books and specific newspapers. In addition to that, we visit museums, and there are lots of stimuli in everyday life.

Do you prefer the analogue or the digital way of working?

sk: I think they go hand in hand. Usually there is an analogue beginning, and the next step we take is digitization. We apply this principle to most of our projects.

ag: Since we can't or don't want to sit at the computer all day, this is a logical decision for us. We make sure that we do some work by hand so that we can transfer everything to another level. We find it more individual if we can produce something which is not determined by a computer programme. Moreover, even computer-aided design can sometimes reach its limits. So the combination of analogue and dialogue activities can open up more possibilities.

You also do a lot of work for magazines like Neon, Kinki, and Zeit Campus. How did that come about?

ag: Actually it was Neon which paved the way for magazines. It happened by several contacts; after all, you know each other in our field of work. This is how it all began …

sk: Well, you know certain people with whom you can begin networking, which is rather important in our line of business. Or you already have several contacts, and that's why the whole thing simply begins to develop. Sometimes we can't even trace it back. But we found out that, in one way or another, there were always some existing connections.

ag: You are represented in a magazine, some people see it, pass it on to other people, and then somebody contacts you. We don't just involve anybody we know, who has to do with our field of work, and then we get an order. It's always a surprise when people contact us, and of course we don't ask them: How did you get the idea to contact us? It just happens automatically.

Are there any projects which have piqued your interest for a long time or have been parked in the drawer for years?

ag: Well … I'm just wondering what we have got in our drawers. *(He laughs)*.

sk: Actually we were always lucky enough to cooperate with our customers, and in this connection we could realize everything we wanted. We were given very much leeway and could always do what we wanted. For this reason we don't have any secret drawer still keeping things in store. We are open to everything. It's exciting to know that any day somebody could drop in and surprise us with interesting projects. So we just wait and see what the future will bring.

ag: There aren't any special projects that we would like to realize at the moment, and actually my drawers are all kept in order, too. *(He laughs)*.

Susanne Kehrer and André Gröger (I LIKE BIRDS) were interviewed by Lynn Blees und Luzia Hein.

240 Glyphen

24h

1440 min

1Tag

! " # $ % & ' ( ) * + , - . / 0 1 2 3

4 5 6 7 8 9 : ; < = > ? @ A B C D E F G

H I J K L M N O P Q R S T U V W X Y Z [

\ ] ^ _ ` a b c d e f g h i j k l m n o

p q r s t u v w x y z { | } ~ ¡ ¢ £ ¤ ¥

¦ § ¨ © ª « ¬ ® ¯ ° ± ´ µ ¶ · ¸ º » ¿ À

Á Â Ã Ä Å Æ Ç È É Ê Ë Ì Í Î Ï Ð Ñ Ò Ó Ô

Õ Ö × Ø Ù Ú Û Ü Ý Þ ß à á â ã ä å æ ç è

é ê ë ì í î ï ð ñ ò ó ô õ ö ÷ ø ù ú û ü

ý þ ÿ ı Ł ł Œ œ Š š Ÿ Ž ž ƒ ˆ ˇ ˘ ˙ ˚ ˛

˜ ˝ ↗ ℗ µ π – — ‘ ’ ‚ “ ” „ † ‡ • … ‰ ‹

› ⁄ € ™ ℗ ◂ ↗ ⧖ ▸ − √ ⋈ ↖ ≈ ≠ ← → ◊ ﬁ ﬂ

HM TILM

Schrift. Typeface. HM Tilm
Gestalter. Designer. Till Wiedeck & Timm Häneke
Label. Foundry. tillwiedeck.com
Jahr. Year. 2010

Als ein Kurz-Projekt wurde HM Tilm mit der selbstgesetzten zeitlichen Vorgabe von 24 Stunden als komplette Monospace-Schrift entwickelt. Der Name geht auf die Vornamen Till und Timm zurück. Die Schrift besteht aus 240 Glyphen. Ihr Hauptcharakter basiert auf der Kombination einfacher, funktionaler Formen mit außergewöhnlichen, verspielten Details.

As a short-term project HM Tilm, an entire monospace typeface, was designed within a self-set time frame of 24 hours. The name originates from the first names, Till and Timm. The character set features 240 glyphs. Its main character is defined by the combination of simple and functional shapes and unusual playful details.

Noch während seines Studiums machte sich der in Berlin lebende und arbeitende *Till Wiedeck*[1] (*1985) 2008 mit dem Studio »HelloMe«[2] selbstständig. Bis zu seinem Abschluss 2010 an der FH Münster arbeitete er parallel dazu für das »Bureau Mario Lombardo«[3] und »Fons Hickmann m23«.[4] Wiedeck veröffentlicht in Büchern und Magazinen (z.B. »Graphic #16, Type Archive Issue«), hält Vorträge und unterrichtet im Rahmen von Workshops.

Das Studio HelloMe existiert bereits seit fünf Jahren. Betrachtet man euer derzeitiges Portfolio, so wird ersichtlich, dass ihr ein breites Spektrum abdeckt, dabei aber meist ein typografischer Fokus vorherrscht. Wie kann man HelloMe und eure Arbeitsweise beschreiben?

tw: HelloMe beschäftigt sich sehr stark mit visuellen Systemen, also dem systematischen Aufbau von Brandings, Strategien und einzelnen Medien. Der Schwerpunkt liegt hier häufig auf Typografie. Sie ist oftmals wichtigster Baustein und auch in Bezug auf Sprache und Kommunikation stetig von Interesse für uns.

Woher stammt das Interesse für Typografie? Hängt es biografisch mit dem Graffiti-Hintergrund zusammen oder hat es sich während des Studiums entwickelt?

Mein erster bewusster Zugang zu Schrift und Typografie erfolgte im Alter von sieben Jahren, als ich eine Dokumentation über Graffiti im Fernsehen sah. Obwohl ich auch zu dieser Zeit schon viel gemalt habe, hat sich mein Interesse für Graffiti in den Jahren danach entwickelt und meine Schulzeit begleitet. Der Fokus hat sich mit Beginn meines Studiums hin zum Grafik-Design und zur Typografie, also der Auseinandersetzung mit Buchstabenformen, verschoben. Heute ist das Thema Graffiti zwar immer noch präsent, betrifft aber eher Aspekte wie Form und Farbe, während meine typografischen Arbeiten hiervon nur insofern betroffen sind, als dass ich ein gewisses Gefühl für die perfekte Form eines Buchstabens durch Graffiti entwickelt habe.

Die Ausstellung »Call for Type« präsentiert die Schrift HM Tilm, die in Zusammenarbeit mit Timm Häneke[5] entstanden ist. In 24 Stunden sind 240 Glyphen erarbeitet worden. Wie hat sich dieses ungewöhnliche Konzept ergeben und wie kam es zur Zusammenarbeit mit Timm Häneke?

Timm habe ich zu Beginn meiner Zeit in Berlin durch gemeinsame Freunde kennengelernt. Als Timm eine Wohnung gesucht hat, ist er unter anderem auch kurzzeitig bei mir untergekommen. Da wir recht gut miteinander ausgekommen sind, entstand die Idee zu einem kurzen, selbst initiierten Projekt. Da wir uns sehr für Typografie interessieren, entstand der Gedanke, innerhalb eines Tages eine Schrift zu schneiden. Letztendlich sind dann innerhalb von 24 Stunden, aufgeteilt auf zwei Arbeitstage, 240 Glyphen entstanden.

1 tillwiedeck.com

2 HelloMe ist ein Berliner Design-Studio mit Schwerpunkt Art Direction, Grafik-Design und Typografie.

3 Das 2004 von Mario Lombardo gegründete Berliner Design-Studio Bureau Mario Lombardo arbeitet im kulturellen Kontext von Kunst, Mode, Musik, Fotografie, Design und Architektur.

4 Fons Hickmann m23 wurde 2001 in Berlin gegründet und wird von Bjoern Wolf und Fons Hickmann geleitet. Das Design-Studio konzentriert sich auf die Gestaltung komplexer Kommunikationssysteme und arbeitet hauptsächlich im kulturellen Bereich.

5 Der aus Frankfurt stammende Timm Häneke studierte an der Akademie U5 in München. Er lebt und arbeitet heute in Berlin. Nach einem Praktikum und Tätigkeit im HORT und im Studio Mario Lombardo arbeitet er derzeit mit und für Kunden aus dem Kultur-, Kunst, Mode- und Verlagsbereich. Zusammen mit William Davis und Wim Michels entwickelt er aktuell die Plattform Szenarios.eu, die sich mit Fragen der Autorschaft sowie politischen und sozialen Folgen von Gestaltung auseinandersetzt und dazu Stellung beziehen möchte.

Vergleicht man diese Herangehensweise mit der Entstehung eines klassischen Schriftentwurfs, werden deutliche Unterschiede ersichtlich. Welche Besonderheiten und Vorteile haben sich bei der HM Tilm ergeben?

Natürlich sind die Glyphen der HM Tilm nicht bis ins Detail ausgefeilt. Den typografischen Feingeist, den wir eigentlich mit Typografie verbinden, kann die Schrift sicherlich nicht vertreten oder deutlich machen. Vielmehr ist die Tilm eine Konzeptarbeit, die sich (ganz simpel) mit der Frage des Machbaren auseinandersetzt. Wir haben eigentlich damit gerechnet zu scheitern. Umso überraschter waren wir, letztendlich 240 Glyphen gezeichnet zu haben. Dadurch, dass es sich um eine Monospace-Schrift handelt, fallen Prozesse wie das Kerning und die Auseinandersetzung mit verschiedenen Parametern nach dem Entwurf der Glyphen größtenteils weg, so dass die Ausarbeitung der grundsätzlichen Idee im Vordergrund steht. Die Schrift selbst bewegt sich zwischen spielerischen Elementen und einer gewissen Stringenz. Es liegt in der Natur der Sache, dass einzelne Buchstabenformen weniger perfekt ausgearbeitet sind als das bei einem klassischen Schriftentwurf üblich ist.

Gibt es Details, die euch heute stören und die ihr im Nachhinein verändern würdet?

Wir sind überzeugt, dass die Schrift an sich Bestand hat. Eine Weiterentwicklung zu einem späteren Zeitpunkt würde das Konzept banalisieren. Die einzige Änderung, die im Nachhinein vorgenommen wurde, war die Ausbesserung an einer Glyphe, die aber das Schriftbild oder den Charakter der Schrift in keinster Weise beeinflusst hat.

Wie fühlt es sich an, die Schrift aus der Hand zu geben und wo stößt die Schrift an ihre Grenzen?

Zunächst einmal hat die Schrift einen relativ kleinen Einsatzbereich, da es sich um eine Monospaced mit nur einem Schnitt handelt. Generell ist die Tilm für Fließtext in mäßigen Längen gedacht. Vor Kurzem habe ich das erste Mal jemanden getroffen, der die Schrift benutzt, aber nicht gekauft hat. Die Schrift wurde für ein persönliches Logo verwendet. Obwohl ich nicht gerade froh darüber bin, wenn Schriften ungefragt verwendet werden, war es mir in diesem Fall relativ egal, da es interessant war zu sehen, wie die Schrift gewisse Grenzen ihres Einsatzbereiches überschritten hat.

Momentan arbeitet HelloMe an der HM June Grotesk,[6] die in abgewandelter Form auch für kommerzielle Projekte benutzt wurde. Wie wichtig ist es für euch, eigene Schriften einzusetzen?

Im Grunde genommen würde ich gerne für jedes Projekt eine eigene Schrift entwerfen. Leider ist dies zeitlich nicht immer möglich. Dennoch versuchen wir diese Idee immer strikter zu verfolgen. Die June, die vor zwei Jahren als freies Projekt entstanden ist, dient häufig als typografische Basis und wird dann projektspezifisch verändert und interpretiert. Für das Branding der Band »The Hundred in the Hands« nimmt die Schrift aufgrund der Veränderung des typografischen Konzepts einen Display-Charakter an. Für Blom&Blom wurde die Logomarke auf Basis der June entwickelt. Außerdem wurden einige Glyphen in Anlehnung an einen Glühdraht gestaltet und im Fließtext eingesetzt.

Die HM Tilm wurde unter anderem in der 16. Ausgabe der koreanischen »Graphic« vorgestellt, die ähnlich wie die Ausstellung »Call for Type« einen Überblick über den zeitgenössischen Schriftentwurf bietet und typografische Trends aufzeigt. Bist Du der Meinung, dass sich zeitgemäßes Grafik-Design und zeitgemäßer Schriftentwurf gegenseitig beeinflussen? Kann man überhaupt von typografischen Trends sprechen?

Sicherlich kann man von typografischen Trends sprechen, aber diese interessieren mich relativ wenig. Vielmehr versuchen wir die richtige Form zu finden, ohne zu adaptieren, was gerade Trend ist. Die »Graphic« hat meines

6 Die HM June Grotesk ist eine serifenlose Schrift, die im Juni 2011 von HelloMe entworfen wurde und momentan von dem Berliner Studio weiterentwickelt und ausgebaut wird.

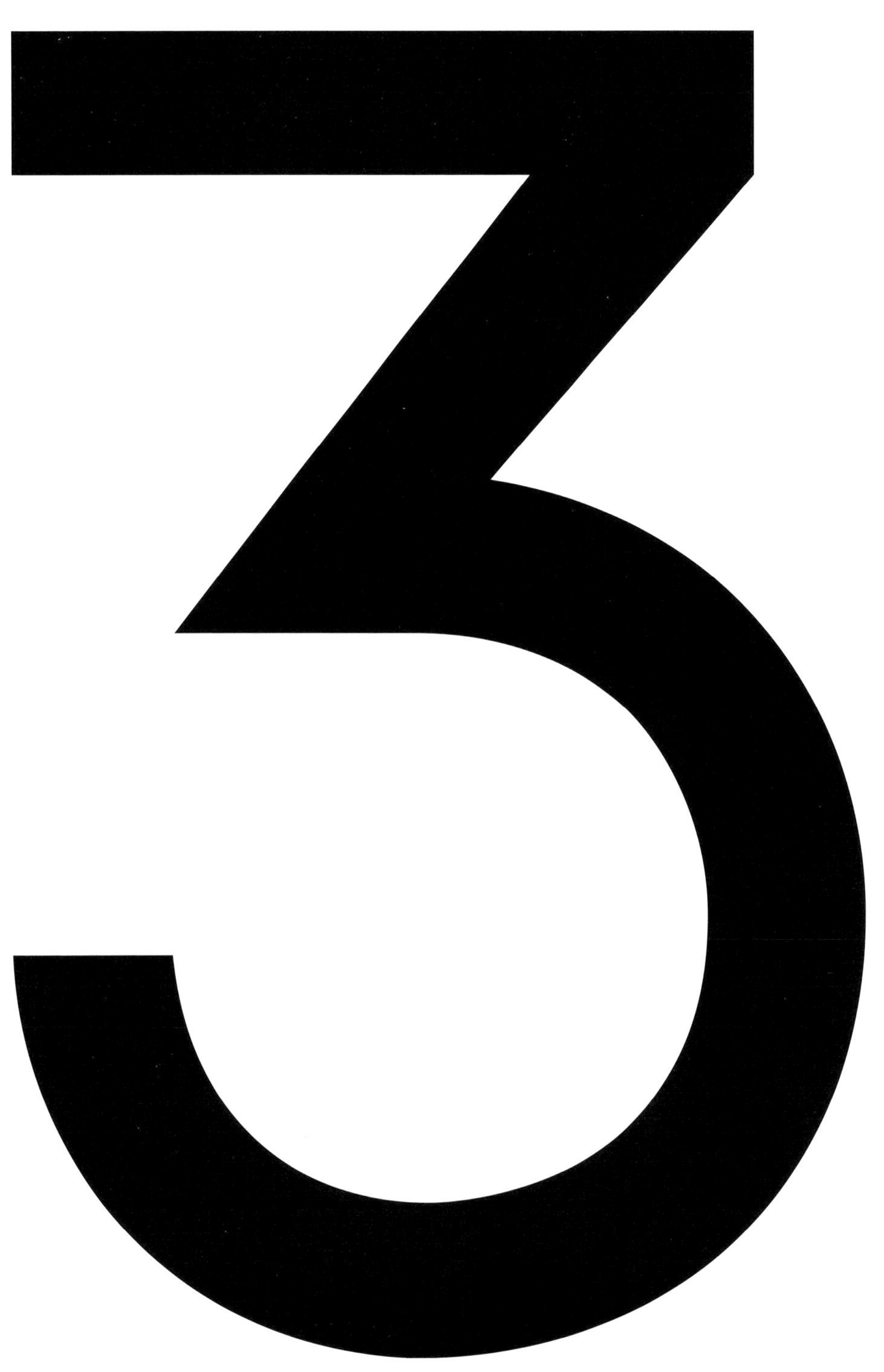

Erachtens seinerzeit ein sehr gutes Kompendium geschaffen, das eine Auswahl an intelligenten Schriften präsentiert. Der Unterschied zwischen zeitgenössischer Typografie und zeitgenössischem Grafik-Design liegt vor allem in dem Zeitaufwand und der nötigen Perfektion, die ein Schriftentwurf erfordert. Während Grafik-Design heute sehr stark durch neue Medien beeinflusst und auch schneller zu erlernen ist, handelt es sich bei Schriftgestaltung immer noch um eine sehr spezielle Disziplin.

Du sprichst neue Medien und damit den digitalen Fortschritt an. Würdest Du sagen, dass hierdurch ein neues Interesse an Typografie entstanden ist? Wo liegen die Vor- und Nachteile dieser Entwicklung? Leidet die Qualität des typografischen Spektrums durch die Quantität?

Grundsätzlich glaube ich nicht, dass die Qualität durch Quantität leidet – das Gegenteil ist der Fall. Natürlich ist nicht jeder Schriftentwurf spannend oder von hoher Qualität. Aber es ist extrem wichtig zu experimentieren und Neues zu erproben und zu lernen. Ich selber befinde mich in diesem Prozess wie viele andere auch. Viel wichtiger ist es, zu reflektieren, wann eine Schrift soweit ist, dass sie der Öffentlichkeit präsentiert werden kann. Hier liegt der eigentliche Grund für eine scheinbare Qualitätsminderung. Zugleich muss man unterscheiden zwischen ernst gemeintem Schriftentwurf und Display-Schriften, die beispielsweise für ein Poster entwickelt werden.

Aus welcher Motivation entstehen bei HelloMe nicht kommerzielle Schriftentwürfe wie Tilm und June Grotesk? Gibt es besondere Inspirationsquellen oder Anregungen für einen Schriftentwurf?

Während die Tilm sich eigentlich während des Zeichnens entwickelt hat und auf einem konzeptionellen Gedanken beruht, hat die June Grotesk auch optische Vorbilder, die weit in die Vergangenheit zurückreichen. Beispielsweise ist der Einfluss durch die Mercator,[7] einem alten Schriftentwurf aus den Niederlanden, erkennbar. Parallel zu diesen Einflüssen aus früheren Zeiten haben wir versucht, einen starken eigenen Charakter herauszuarbeiten, der sich durch eine gewisse Freundlichkeit und verschmitzte Details äußert.

Man kann also sagen, dass auch der Prozess des Schriftentwurfs Inspirationsquellen für den Entwurf birgt?

Auf jeden Fall. Ich würde auch immer noch behaupten, dass die June Grotesk in keiner Weise vollendet ist. Die meisten Anfragen bezüglich der Schrift lehnen wir ab, weil der Prozess noch nicht abgeschlossen ist. Es wird noch ein bis zwei Jahre dauern, bis die Schrift ausgebaut ist und der Öffentlichkeit zugänglich gemacht werden kann.

An welchen Projekten mit typografischem Schwerpunkt arbeitet HelloMe im Augenblick?

Derzeit arbeiten wir am »Berlin Art Prize«, für den wir das Logo geschnitten haben, und am Ausbau der Schrift für den Katalog. Da es sich hierbei um ein jährlich wiederkehrendes Projekt handelt, wird der Schriftentwurf sicherlich auch nach Erscheinen des ersten Katalogs von Relevanz sein und über die kommenden Jahre weiter entwickelt werden.

Das Interview mit Till Wiedeck führten Robin Scholz und Felix Rank.

7 Die Mercator wurde 1959 von Dick Dooijes und G. W. Ovink für die Amsterdam Type Foundry entwickelt.

Till Wiedeck[1] (*1985), who lives in Berlin, started his own Studio »HelloMe«[2] in 2008 while still studying. Until his graduation 2010 at the University of Applied Sciences in Münster, he concurrently worked for the »Bureau Mario Lombardo«[3] and »Fons Hickmann m23«.[4] Wiedeck publishes in Books and Magazines (e.g. »Graphic #16, Type Archive Issue«), he lectures and teaches at workshops.

The studio HelloMe is already five years old. A look at your current portfolio shows that you cover a wide range of design services, which, however, mainly focus on typography. How would you describe HelloMe and your way of working?

tw: HelloMe is busily engaged in visual systems, this means in the systematic development of branding, strategies, and single media. There the focus is frequently on typography. Typography is often the most important element, and we are always interested in it in terms of language and communication, too.

How did you get interested in typography? Has this to do with your biographical background and your enthusiasm for graffiti, or did it develop during your studies?

When I was seven, I watched a TV documentary on graffiti, and this was my first conscious approach to script and typography. Though I already painted a lot at that time, my interest in graffiti began to develop from then on, and I never stopped to be interested in it in my schooldays. When I began my studies, I shifted my focus to graphic design and typography, and so I dealt with the subject of letter forms. Today I'm still interested in the subject of graffiti. However, it rather concerns such aspects as shapes and colours. My typographic work is only influenced by it to such a degree as graffiti helped me to develop a certain feel for perfect letter forms.

The »Call for Type« exhibition shows the HM Tilm font, which was created in collaboration with Timm Häneke. 240 glyphs were developed in 24 hours. How did it happen that you had such an unusual concept and that you cooperated with Timm Häneke[5]?

At the beginning of my time in Berlin, Timm and I got to know each other because we had mutual friends. When Timm was looking for an apartment, he also stayed at my home for a while. Since we got along rather well, we got the idea for a short project of our own. We are both very much interested in typography. For this reason it occurred to us that we could cut a font within a day. Finally we created 240 glyphs within 24 hours or actually two workdays.

Compared with a traditional design of a font, your approach is clearly different. What are the characteristics and advantages of HM Tilm?

Of course the glyphs of HM Tilm weren't designed down to the smallest detail. Certainly this font cannot represent or reveal a sophisticated font design which we actually associate with typography. In fact, Tilm can be considered a conceptual work which (quite simply) wants to find out what's feasible. Actually we thought that we might fail. So we felt all the more surprised by the fact that we had succeeded in drawing 240 glyphs. Since it's a monospaced font, several processes like kerning and dealing with diverse parameters could be, for the most part, omitted after the glyphs had been designed, so that the focus was actually put on the elaboration of the basic idea. The font itself ranges between playful elements and a certain kind of rigour. It's in the nature of things that the single letter forms are less perfectly elaborated than it would be the case for a traditional typeface design.

Are there any details which you dislike today and which you would like to change with the benefit of hindsight?

We are convinced that this font is something enduring. If there were any further development of it in the future, this would banalize its concept. The correction of a single glyph was the only thing which we changed afterwards. However, this didn't have any influence at all on the typographic design or the characteristics of this font.

How do you feel about letting this font go out of your hands, and where does this font reach its limits?

To begin with, this font has a relatively small range of use because it's a monospaced font with only one cut. Tilm is generally intended for running text of moderate lengths. Recently I met for the first time somebody who employed this font without having bought it. He used the font for a personal logo. Though I'm usually not very happy when any of my fonts are used without permission, I could more or less accept it in this special case because it was interesting to see how the font had exceeded certain boundaries of its actual field of application.

At the moment HelloMe works on HM June Grotesk,[6] which was also used in a modified form for commercial projects. How important is it to you to use your own fonts?

Actually I would like to develop a special font for each project. Unfortunately, due to time constraints, this is not always possible. However, we increasingly

1 tillwiedeck.com

2 HelloMe is a Berlin design studio with emphasis on art direction, graphic design and typography.

3 The Berlin Design Studio Mario Lombardo, established in 2004 by Mario Lombardo, works in the cultural context of art, fashion, music, photography, design and architecture.

4 Fons Hickmann m23 was founded in Berlin in 2001 and is run by Bjoern Wolf and Fons Hickmann. The design studio is focusing on the design of complex communication systems and mainly works in the cultural segment.

5 Frankfurt-born Timm Häneke studied at the Akademie U5 in Munich and is now living and working in Berlin. After an internship and a period of working for HORT and for Studio Mario Lombardi, he now works for and with clients from the cultural, art, fashion and publishing sector. Together with William Davis and Wim Michels he is currently developing the platform Szenarios.eu, which will discuss and take a stance in questions of authorship, as well as political and social results of design.

6 The HM June Grotesk is a sans-serif typeface designed in 2011 by HelloMe and is currently further developed by the Berlin studio.

try to pursue this idea. June, which was created as an independent project two years ago, frequently serves as typographic basis and is then modified and interpreted according to the requirements of the individual project. So the typographic concept of the font was changed for the branding of the »The Hundred in the Hands« music duo, and then the font resembled a display typeface. The logo mark for Blom&Blom was also developed on the base of June. In addition to that, several glyphs, which had been designed in the style of light bulb filaments, were inserted into running text.

HM Tilm was also presented in the 16th edition of the Korean »Graphic«, which, similar to the »Call for Type« exhibition, gives an overview of contemporary type design and shows typographic trends. Do you think that modern graphic design and contemporary type design influence each other? Can you actually speak of typographic trends at all?

You can certainly speak of typographic trends. However, I'm hardly interested in them. In fact, we try to find the right form without any adaptation, which has become trendy. I'm of the opinion that »Graphic« created an excellent compendium with a selection of intelligent fonts at that time. The difference between contemporary typography and modern graphic design lies mainly in the expenditure of time and the perfection which the design of a font requires. Whereas today's graphic design is highly influenced by new media and is also easier to learn, type design is still a pretty specialized discipline.

You are talking about new media and so you are referring to digital progress. Do you think that this development has led to a new interest in typography? What are the advantages and disadvantages of this development? Does high quantity have a negative impact on the quality of the typographic spectrum?

Basically, I don't think that more quantity means less quality. – It's quite the opposite. Of course not every design of a font is captivating or has a high level of quality. But it's extremely important to experiment and to try and learn something new. That's exactly the process I am involved in at the moment, like so many others. In fact, it's far more important to know when a font is ready for its presentation to the public. If there seems to be a reduction in quality, it can actually be explained by this fact. At the same time, you must differentiate between »serious« fonts and display typefaces, which were developed for posters, for instance.

What's your motivation at HelloMe for creating non-commercial fonts like Tilm and June Grotesk? Have you got any particular sources of inspiration or stimuli for the development of a font?

Whereas Tilm has actually been developed during the drawing of it and is based on a conceptual idea, June Grotesk also followed some visual examples which go far back in time. So, for instance, it can be noticed that the font was influenced by Mercator,[7] an old typeface from the Netherlands. Parallel to such influences from past times, we tried to elaborate some strongly individual attributes which are characterized by a certain friendliness and some playful details.

Does this mean that the process of creating a font can also serve as source of inspiration for its design?

Absolutely! I'm also standing by my opinion that June Grotesk is still far from perfect. When we are asked for this font, we mostly refuse such a request because the process of this font hasn't been finished yet. It will take us another year or two to develop the font in such a way that it can be made available to the public.

What projects with focus on typography is HelloMe currently working on?

At present we're working for the »Berlin Art Prize«. We cut its logo and we are working on the completion of the font for the respective catalogue. Since this work is done for an annual project, the font will presumably be used even after the publication of the first catalogue, and it will be further developed in coming years.

Till Wiedeck was interviewed by Robin Scholz and Felix Rank.

7 The Mercator was developed in 1959 by Dick Dooijes and G.W. Ovink for the Amsterdam Type Foundry.

Berlin is like being abroad in Germany.

Karbid Display Pro Bold Italic – 36pt

A disgusting city, this Berlin, a place where no one believes in anything.

Karbid Slab Pro Bold – 18pt

And now we come to the most lurid Underworld of all cities – that of post-war Berlin. Ever since the declaration of peace, Berlin found its outlet in the wildest dissipation imaginable. The German is gross in his immorality, he likes his *Halb-Welt* or underworld pleasures to be devoid of any *Kultur* or refinement, he enjoys obscenity in a form which even the Parisian would not tolerate.

Karbid Text Pro Regular, Italic – 6pt

Ihr Völker der Welt, ihr Völker in Amerika, in England, in Frankreich, in Italien! Schaut auf diese Stadt und erkennt, dass ihr diese Stadt und dieses Volk nicht preisgeben dürft und nicht preisgeben könnt!

Karbid Pro Medium – 9pt

Karbid Display Pro Light, *Italic*
Karbid Display Pro Regular, *Italic*
Karbid Display Pro Medium, *Italic*
Karbid Display Pro Bold, *Italic*
Karbid Display Pro Black, *Italic*

Karbid Pro Light, *Italic*
Karbid Pro Regular, *Italic*
Karbid Pro Medium, *Italic*
Karbid Pro Bold, *Italic*
Karbid Pro Black, *Italic*

Karbid Text Pro Light, *Italic*
Karbid Text Pro Regular, *Italic*
Karbid Text Pro Medium, *Italic*
Karbid Text Pro Bold, *Italic*
Karbid Text Pro Black, *Italic*

Karbid Slab Light, *Italic*
Karbid Slab Regular, *Italic*
Karbid Slab Medium, *Italic*
Karbid Slab Bold, *Italic*
Karbid Slab Black, *Italic*

Schrift. Typeface.	Karbid Pro
Gestalter. Designer.	Verena Gerlach
Label. Foundry.	fraugerlach.de
Jahr. Year.	1998

Die Elemente der Karbid gehen zurück auf die Anzeigen- und Fassadenbeschriftung um 1900, von der durch den Wiederaufbau Berlins nach dem Mauerfall vieles verlorengegangen ist. Die Schrift möchte jedoch nicht als Wiederbelebung wahrgenommen werden und verzichtet auf nostalgischen Glanz; vielmehr versucht sie eine neue Interpretation der Originale im Licht moderner Designpraktiken. Karbid-Text ist auf das absolut Notwendige einer Schrift reduziert, wodurch sie sehr lesbar wird, besonders in den kleineren Punktgrößen.

Trotz dieser Reduktion ist Karbid-Text eine Schrift, die durch ihre Lebendigkeit besticht. Die Bögen, die die Serifen ersetzen und die Zeichen miteinander verbinden, erzeugen den Eindruck von fließender Bewegung.

The elements that make up Karbid hark back to original sign and façade lettering at the turn of the last century, much of which has been lost in the course of the ongoing redevelopment of Berlin after the fall of the wall. Not wanting to be seen as a mere revival, the font avoids nostalgic gloss, but attempts a fresh interpretation of the originals in the light of modern design principles.

Karbid-Text has been trimmed down to the bare essentials of a text face which makes it eminently readable, especially at small point sizes. Despite this back to basics reduction, Karbid-Text is a font that captivates through its sheer liveliness. The sweeps that replace the serifs and link the characters create a flowing movement.

Hellersdorf

Görlitz

Schöneberg

Tiergarten

PANKOW

Kreuzberg

KARBID PRO

Karbid Display Pro Bold
Karbid Slab Pro Bold

Karbid Pro Italic
Karbid Pro Bold

Karbid Display Pro Light
Karbid Text Pro Black

Die gebürtige Berlinerin *Verena Gerlach*[1] (*1971) studierte von 1993 bis 1998 an der Kunsthochschule Berlin Weißensee Visuelle Kommunikation. Mit »fraugerlach« gründete sie ein Büro für Grafik-Design, Type Design und Typografie. Meist sind es typografische Fundstücke wie alte Stempel, Straßenschilder und Plastikbuchstaben, die sie zu eigenen Schriften inspirieren. So wurde die FF Karbid, eine ihrer bekanntesten Schriften, einer alten Berliner Fassadenbeschriftung entlehnt. Die FF Karbid wurde 2011 redesigned und unter dem Namen *FF Karbid Pro* erneut bei »FontFont« veröffentlicht. Weitere Schriften von Gerlach sind die LT Pide Nashi und die EF Aranea, die PTL Touja Sans sowie die PTL Tephe. Verena Gerlach unterrichtet an internationalen Hochschulen.

Frau Gerlach, wie sieht denn Ihr Arbeitsalltag aus? Wie beginnen Sie Ihren Tag?

vg: Wenn es zeitlich passt, versuche ich zur Arbeit zu laufen – hier am Kanal ist das sehr schön. Dann fange ich um neun an. Ab und an beantworte ich die ersten Mails schon gegen acht Uhr von zu Hause aus. Das ist leider Alltag. Gegen halb eins mache ich Mittagspause. Dann geht das ungefähr bis sieben, acht, neun, zehn. Wenn ich es schaffe, arbeite ich ganz gerne abends noch zu Hause.

Woher kam Ihr Interesse am Type Design? Kam das schon vor dem Studium, währenddessen oder doch erst danach?

An Typografie hatte ich schon immer Interesse. Das kann ich ganz klar sagen. Das bestand bereits im Kindergarten. Meine Kinderzeichnungen waren schon modulartig aufgebaut. Mein Arbeitsalltag ist zurzeit nicht Type Design. Ich bekomme das zeitlich nicht hin. Gerade mache ich fast ausschließlich Buchgestaltung, Grafik-Design und gebe Unterricht.

Wenn Sie eine Schrift entwickeln, machen Sie das nebenbei?

Die Schriften, die ich herausgebracht habe, machte ich in Vollzeit. Ich nehme mir dann wochenlang frei.

Dann nehmen Sie keine Aufträge mehr an und arbeiten ausschließlich an der Entwicklung einer neuen Schrift?

Genau, anders bekomme ich das nicht hin. Ich muss täglich mindestens sechs Stunden daran sitzen, nur um reinzukommen. Schriften brauchen wirklich viel Zeit. Das kann man nicht nach Feierabend machen.

Um auf Ihre vorgestellte Schrift, die Karbid, zu kommen. Was hat Sie zu dieser Schrift bewegt und was waren Ihre Inspirationsquellen?

Das sind die Ladenbeschriftungen aus Ost-Berlin und Mitte – eigentlich aus ganz Berlin. Im Prenzlauer Berg und Mitte haben sie die DDR überlebt. Ich habe sehr viele Fotos gemacht und gesammelt. Das war meine Grundidee für das Design; ich bin der Sache wirklich auf den Grund gegangen.

1 fraugerlach.de

Wie kam Ihr erster Gedanke dazu, wobei entstand die Idee zur Schrift?

Das war ein Riesenprojekt. Ich mache das schon seit zwei Dekaden. In den Achtzigern war ich – was wir West-Berliner ja konnten – in Ost-Berlin unterwegs. Schon damals sind mir die Ladenbeschriftungen aufgefallen. Beim Mauerfall habe ich mir dann meine Kamera gegriffen und alles fotografiert. Bis zum dritten Studienjahr wusste ich nicht, dass man Buchstaben eigens als Font entwickeln kann.

In einem Kurs, den ich bei Matthias Gubig[2] und de Groot[3] belegt hatte – ich glaube, wir waren die erste Klasse von Lucas in Deutschland – hatte ich dann das Thema »Neukölln«. Das war für mich die Inspiration. Ich habe eine Mischung aus Fraktur und Arabisch ausprobiert. Wir arbeiteten noch mit Fontographer und sogar Font Studio. Da habe ich erst einmal verstanden, dass man Schriften tatsächlich selbst machen kann.

Hatten Sie Hilfe bei der Karbid?

Ja, das war meine Diplomarbeit. Sie wurde damals betreut von Matthias Gubig und Lucas de Groot, der damals Gastdozent war. Dann wechselte ich zu Ole Schäfer,[4] der meine Schrift in der Endphase betreute. 2011 habe ich sie redesigned.

Zur Karbid Pro?

Ja, ich habe sie überarbeitet und runder gemacht. Mit Fred Smeijers,[5] der die Karbid noch aus ihrer Ursprungszeit kannte, hatte ich mich kurzgeschlossen. Ich habe sie um Italics und Slabs ergänzt und auch diverse OpenType-Features eingebaut.

Würden Sie gerne noch weitere Schriften entwickeln?

Klar. Wenn ich mal wieder Zeit und Muße habe, kommt garantiert noch eine.

Wenn Sie die Karbid in einer Publikation sehen, was geht da in Ihnen vor?

Ich freue mich. Ich finde das eigentlich immer toll. Wenn die Karbid nicht so schön eingesetzt ist, fühle ich mich wie Eltern, deren Kinder die falsche Musik hören.

Das haben Sie schon erlebt?

Na klar. Auch mit den anderen Schriften. Damit muss man leben. Man hat die Schrift zwar entwickelt, aber nicht das Design, das damit gemacht wurde. Man kann es mögen oder nicht mögen. Meistens freue ich mich total. Manchmal bin ich auch überrascht und denke: »Wow, so hätte ich sie gar nicht eingesetzt.«

Was gefällt Ihnen an der Karbid am besten? Haben Sie einen Lieblingsbuchstaben?

Ich bin ein großer Fan vom kleinen a in Slab. Dem doppelstöckigen a. Das finde ich super. Ach, da gibt es einige. Jeden Tag mag ich einen anderen lieber. Heute mag ich das kleine a.

Haben Sie sich auch an etwas satt gesehen?

Ich mag die Urkarbid nicht mehr.

Haben Sie Vorbilder?

Es gibt Leute, die ich total super finde und die mich inspirieren. Die können schon lange tot

2 Der Buchgestalter und Grafiker Matthias Gubig (*1942) wurde mehrfach in den Wettbewerben »Die 100 besten Plakate«, »Schönste Bücher aus aller Welt« und »Die schonsten Bucher der DDR« ausgezeichnet.

3 Die immens ausgebaute Schriftfamilie FF Thesis des niederländischen Schriftentwerfers und Typografen Lucas de Groot (*1963) wurde zum Meilenstein dessen, was an Systematik und Perfektion erreichbar ist. 2000 gründete Lucas de Groot die Type Foundry »LucasFonts«. De Groot gestaltete u.a. die Hausschriften des Nachrichtenmagazins »Der Spiegel«, des Fernsehsenders »ARD«, der SPD und der deutschen Tageszeitung »taz«.

4 Ole Schäfer (* 1970) ist ein deutscher typografischer Gestalter und Schriftentwerfer. Er war involviert in den Ausbau der ITC-Schrift Officina, der FF Meta sowie der FF Info. 2000 veröffentlichte er seine FF Fago, zwei Jahre später gründete er den Schriftenverlag und -dienstleister Primetype (→ S. 180).

5 Der niederländische Schriftdesigner und Autor Fred Smeijers (*1961) arbeitet schwerpunktmäßig in der typografischen Forschung und Entwicklung für Produkthersteller. Smeijers entwarf zahlreiche Schriftfamilien, u.a. die Arnhem. Er ist Mitbegründer und Creative Director von OurType und Autor der Bücher »Counterpunch« und »Type now«.

sein oder noch leben. Das sind ganz viele. Meine Vorbilder sind eigentlich immer die, die nicht nur eine Sache machen. Die den ganzen grafischen Bereich abdecken. Die vom Plakat bis zur Schrift und zum Interior-Design alles machen. Die finde ich meistens am spannendsten.

Können Sie Namen nennen?

Ich bin großer Lucian Bernhard-Fan.[6] Aus der Pop Art finde ich die Sachen von Robert Indiana[7] und Sister Corita[8] (Mary Corita Kent) total klasse. Veronika Burian[9] und José Scaglione[10] machen fantastische Schriften. Ich verehre sie sehr, auch wenn sie weniger interdisziplinär arbeiten.

Wenn man eine eigene Schrift entwickelt hat und deren Konzeption und Besonderheiten kennt, neigt man dann dazu, diese selber häufiger einzusetzen?

Das mache ich nicht. Dabei muss man auch den Mut haben zu sagen, ich nehme sie nicht, da sie nicht passt. Ich würde mich als Buchgestalterin unglaubwürdig machen, wenn ich sie verwenden würde, obwohl sie nicht passt.

Type Design ist sehr von Männern dominiert. Wie empfinden Sie das als Frau?

Ich bin die, die immer Prügel bezieht, da ich das so sehe und natürlich den Mund aufmache und auch darüber schreibe. Wenn jemand behauptet, es sei nicht so und es läge an der Technik, dann gehe ich die Wand hoch. Klar ist das so.

Glauben Sie, dass sich das ändern wird?

Das weiß ich nicht. Das hat nichts mit Type Design zu tun. In Deutschland wird die Situation generell immer problematischer. Wenn ich jetzt höre, unsere Familienministerin tritt zurück, damit sie sich um die Familie kümmern kann, brauche ich dazu eigentlich nichts mehr zu sagen, oder?

Ich sehe Type Design nur als ein weiteres Feld in unserer Gesellschaft. Es gibt viele Berufe, die zeitintensiv sind. Wenn es darum geht, dass sich die Frauen um die Kinder kümmern, dann ist keine Zeit mehr da. Für die Frauen, die ich kenne, die Type Design machen UND Familie haben, ist das heftig. Von denen gibt es ganz, ganz wenige.

Also könnte das Gebiet mehr von Frauen dominiert werden, wenn unsere Gesellschaft nicht so wäre, wie sie ist?

Genau. Also, was heißt dominiert? Das Schönste wäre ja, wenn keiner dominiert.

Wie stellen Sie sich Ihre Zukunft vor?

Nicht viel anders als es jetzt ist. Ich kann mir auf der anderen Seite auch vorstellen, dass ich eventuell etwas komplett anderes machen werde.

Können Sie sich vorstellen, in eine andere Stadt zu ziehen, in eine Stadt, die Ihre Schriften so beeinflusst wie Berlin?

Ich habe es ein paar Mal versucht, Berlin zu verlassen. Allerdings sage ich auch immer, dass ich hier bleiben kann und die Stadt sich verändert. Berlin ist alle drei Jahre anders. Ob man das nun will oder nicht. Es kann gut sein, dass ich irgendwann umziehe. Da bin ich offen. Aber ich habe das Glück, dass ich durch meine Reisen unheimlich viel Inspiration bekomme. Wenn es mir irgendwo gefällt, dann kann ich auch gerne dort bleiben.

Das Interview mit Verena Gerlach führten Julia Heil und Tobias Villmeter.

6 Der deutsche Grafiker und Typograf Lucian Bernhard (1883–1972) gestaltete Plakate für bedeutende Marken wie Stiller, Pelikan, Kaffee Hag, Bosch oder Faber-Castell.

7 Robert Indiana (*1928, ursprünglich Robert Clark) ist ein US-amerikanischer Maler sowie ein Hauptvertreter der Pop Art und der Signalkunst. Bekannt wurde er durch seine plakativen Zeichenbilder, die zu den radikalsten Äußerungen in der Pop Art zählen. Seine Arbeiten bestehen meist aus Zahlen, Buchstaben und fünfstrahligen Sternen.

8 Sister Corita (Mary Corita Kent, 1918–1986) war eine Nonne, die in der katholischen Immaculate Heart Community in Los Angeles lebte und sich gegen soziales Unrecht und den Vietnamkrieg engagiert hat. In Siebdrucktechnik fertigte sie Transparente für Demonstrationen und Plakate. Außerdem organisierte sie Gesprächsreihen und happeningartige Veranstaltungen, die ästhetische mit gesellschaftspolitischen Problemstellungen vereinten.

9 Veronika Burian (→ S. 32)

10 José Scaglione (*1974) ist ein argentinischer Grafik- und Type-Designer, der an den Universitäten in Rosario und Buenos Aires in Argentinen unterrichtet. Darüber hinaus ist er seit 2007 Mitglied des Vorstandes des »Association Typographique Internationale«. Er gründete 2006 zusammen mit Veronika Burian die Type Foundry TypeTogether (→ S. 32, 60).

»Wenn die Karbid nicht so schön eingesetzt ist, fühle ich mich wie Eltern, deren Kinder die falsche Musik hören.«

Verena Gerlach
→ S. 125

From 1993 to 1998, *Verena Gerlach*[1] (*1971 in Berlin) studied visual communication at the Kunsthochschule Berlin Weißensee. She established »fraugerlach«, a studio for graphic design, type design, and typography. Frequently typographic finds, such as old rubber stamps, road signs, and plastic letters, serve as inspiration for Gerlach's type design. So the »FF Karbid«, one of her most famous fonts, was inspired by an old lettering of a Berlin façade. In 2011, her FF Karbid font was redesigned and published by »FontFont« as *Karbid Pro*. Other fonts which were created by Gerlach are LT Pide Nashi, EF Aranea, PTL Touja Sans as well as PTL Tephe. Verena Gerlach teaches at several international universities of applied sciences.

Mrs. Gerlach, how would you describe your typical work day? How do you start your day?

vg: If time permits, I try to walk to work, which is very nice along the canal. Then I start my work at 9 o'clock. Once in a while, somewhere around 8 o'clock, I answer my first e-mails already at home. This is unfortunately daily routine. At around 12:30, I have a lunch break. I finish my work at around seven, eight, nine, or ten o'clock in the evening. If I can find the time, I also like to continue my work at home.

How did your interest in type design develop? Were you already interested in type design before your studies, or did your interest develop during your studies or even later?

I was always interested in typography. This is a fact. Already in kindergarten I was fascinated by it. The drawings I made as a child were already composed of single modules. At the moment, my typical work day doesn't imply type design. I haven't got enough time for that. For the time being, I'm almost exclusively occupied with book design and graphic design, and I am teaching.

When you develop a typeface, can you do this during your usual working time?

It was a full-time job for me to develop my typefaces. In such a case, I take several weeks off.

Then you don't accept any new orders, and you work exclusively on the development of the new font.

Exactly, otherwise it would not be possible for me. It takes me at least six hours a day just to get myself familiar with it. The development of typefaces really takes a lot of time. It's not a job that can be done after work.

Let's talk about Karbid, a typeface which you developed and which is also shown here. What was your motivation for the development of this font? What was your source of inspiration?

The lettering on the shops in East Berlin and in Berlin's Mitte district inspired me, but actually it was the lettering of all Berlin shops. In the area around Prenzlauer Berg and Berlin-Mitte, the lettering survived the German Democratic Republic. I took a great many pictures and collected them. This was the basic idea for its design; I really got to the bottom of it.

How did you get your first idea of typefaces? When did you start to create fonts?

This was a huge project. Actually I have been doing this for two decades. In the 1980s, I was in East Berlin – as a resident of West Berlin I could of course go there. Already in those days I noticed the lettering on the shops. After the fall of the Berlin Wall, I took my camera with me and made pictures of everything. Almost at the end of year three of my studies, I didn't know that letters could be specially created for the production of fonts.

It was in a course by Matthias Gubig[2] and Lucas de Groot[3] – I think we were Lucas' first course in Germany – that I had to deal with the subject of the Neukölln typeface. This was my inspiration. I tried a mixture of Fraktur and Arabic. At that time, we still used Fontographer and even Font-Studio. This was the first time I realized that you can actually create your own fonts.

Did you have any help with the development of Karbid?

Yes, this was my diploma thesis. Matthias Gubig and Lucas de Groot, a visiting professor at that time, were my responsible mentors. Then, when my font was nearly finished, Ole Schäfer[4] became my mentor and contact person with regard to the development of this font. In 2011, I redesigned the font.

1 fraugerlach.de

2 The book designer and graphic artist Matthias Gubig (*1942) repeatedly won prizes in the contests »The 100 best Posters«, »Most Beautiful Books from all over the World« and »The most beautiful Books of the DDR«.

3 The huge type family FF Thesis, developed by the Dutch type designer and typographer Lucas de Groot (*1963) became a milestone for systematics and perfection. In 2000 Lucas de Groot launched the type foundry »Lucas Fonts«. De Groot designed, among others, the typefaces for the Magazine »Der Spiegel«, for the TV station »ARD«, the SPD and the German daily paper »taz«.

4 Ole Schäfer (*1970) is a German typographer and type designer. He was involved in the development of the ITC type Officina, the FF Meta as well as the FF Info. In 2000 he published his FF Fago, two years later he launched his type foundry and service company Primetype (→ S. 185).

5 The Dutch type designer and author Fred Smeijers (*1961) focuses on typographic research and development for producers. Smeijers developed numerous type families, e.g. the Arnhem. He is co-founder and Creative Director of Our Type and author of »Counterpunch« and »Type Now«.

It was redesigned and became Karbid Pro?

Yes, I reworked the typeface and made it rounder. I had got in touch with Fred Smeijers,[5] who knew Karbid from its beginnings. I added Italic and Slabs and also built in several OpenType features.

Would you like to develop some more typefaces?

Yes, of course. When I have enough time and energy, I will definitely develop another one.

What is your reaction when you see Karbid in a publication?

I feel happy. Actually I always think that it's great; and if Karbid is not used in an appropriate way, I feel like parents whose children listen to the wrong music.

Have you already gone through this?

Sure! And it's the same with other fonts. You just have to live with that. You develop a typeface, but you can't control the design for which it is used. You can like or dislike it. Most of the time, I'm absolutely delighted with the results. Sometimes I'm even surprised thinking: »Wow, I would never have thought of using this typeface in such a way«.

Which element of Karbid do you like most? Have you got any favourite letter?

I'm a big fan of the lower case form of the letter »a« in Slab, namely the double-storey »a«. I think it is really great. Well, I have several favourite letters. Every day I prefer another one. Today, for instance, I like the small letter »a«.

Is there anything you are fed up with?

I don't like the first Karbid anymore.

Have you got any role models?

There are people who are really super and inspire me. Some of them have long been dead, others are still alive. There are a great many of them. Actually my role models are those people who don't focus on one thing, but cover the complete graphic spectrum ranging from posters and typefaces to interior design. They are the most fascinating for me.

Could you give us some names?

I'm a big fan of Lucian Bernhard.[6] In the field of Pop Art, I'm totally enthusiastic about the work done by Robert Indiana[7] and Sister Corita[8] (Mary Corita Kent). Veronika Burian[9] and José Scaglione[10] create fantastic fonts. I really admire these persons though they do not work in such an interdisciplinary way.

If somebody develops his own typeface and knows about its purpose and use, is he also prone to use it frequently in his own publications?

I don't do that. You must be honest enough to decide against it, if it doesn't work. If I used my font thinking it was inappropriate, I would lose my credibility as book designer.

Type Design is a male-dominated field. How do you feel about this as a woman?

I am the one who always takes punches because I can clearly see the situation as it is. Certainly I speak up for an improvement and of course also write about it. If anybody denies this fact and says that it is all up to technical reasons, I go off. It's quite evident that the situation is like it is.

Do you think that this situation will change?

I don't know. It has not got anything to do with type design. In Germany the whole situation is getting more and more difficult. When I hear that our Federal Minister for Family Affairs wants to resign so that she can take care of her family … There is no need to say anything else concerning this matter …

Type design is only one of these fields in our society. There are many other jobs which are time-consuming. If women are supposed to take care of children, there is no more time left. It is very difficult for some women I know to design type AND have a family. There are only a few of them.

Do you mean that this field would be more dominated by women if our society was different?

Exactly! But why »dominate«? Of course it would be better not having any domination at all.

How do you imagine your future?

Not very different than it is now. However, I can imagine that I could possibly do something completely new.

Could you imagine moving to a different city, a city which would influence your typefaces in the same way as Berlin did?

I have tried several times to leave Berlin. However, I always say that I can stay here – it's the city which is changing. Every three years, Berlin is different, whether you like it or not. It is possible that I will move to another place someday. I am open to that. But I fortunately get a great deal of inspiration during my travels. I can easily stay somewhere else, if I like the place.

Verena Gerlach was interviewed by Julia Heil and Tobias Villmeter.

6 The German graphic artist and type designer Lucian Bernhard (1883–1972) designed posters for important brands, e.g. Stiller, Pelikan, Kaffee Hag, Bosch or Faber-Castell.

7 Robert Indiana (originally Robert Clark, *1928) is an American painter as well as one of the main protagonists of Pop Art and Signal Art. He became well-known through his striking paintings which are among the most radical manifestations in Pop Art. His work typically consists of numbers, letters and five-point stars.

8 Sister Corita (Mary Corita Kent, 1918–1986) was a nun of the Roman-Catholic Community of the Immaculate Heart, Los Angeles, who campaigned against social injustice and the Vietnam War. She produced banners for demonstrations in screen print technique and organised talks and happenings, combining aesthetic and socio-political problems.

9 Veronika Burian (→ S. 37)

10 José Scaglione (*1974) is an Argentinian graphic and type designer teaching at the Universities of Rosario and Buenos Aires. Since 2007 he has also been a Board Member of the Association Typographique Internationale. In 2006 he founded, together with Veronika Burian, the type foundry »TypeTogether« (→ S. 37, 63).

PETER BIL'AK & PIETER VAN ROSMALEN

Pretty, Boring, Crazy.

Schrift. Typeface. Karloff
Gestalter. Designer. Peter Bil'ak & Pieter van Rosmalen
Label. Foundry. typotheque.com
Jahr. Year. 2012

Karloff lotet die Möglichkeiten unvereinbarer Unterschiede aus, d.h. sie widmet sich der Frage, wie zwei Extreme zu einem kohärenten Ganzen vereint werden können.

Zu Beginn wurden die hoch kontrastierenden Didot-Schriften analysiert, die von vielen Fachleuten als zu den Schönsten gehörend betrachtet werden und die exzentrische Italian, die ebenfalls geschaffen wurde, um die Aufmerksamkeit der Leser zu wecken, weil sie anders ist als erwartet. Keine andere Schrift in der Geschichte der Typografie hat zu ähnlich negativen Reaktionen geführt wie die Italian.

Karloff, das Ergebnis dieses Projekts, verbindet die hoch kontrastierende moderne Schrift von Bodoni und Didot mit der monströsen Italian. Der Unterschied zwischen der attraktiven und der irritierenden Schrift wird von einem Design-Parameter bestimmt, nämlich dem Kontrast zwischen fett und dünn. Nachdem zwei völlig gegensätzliche Versionen entworfen wurden, ergab ein genetisches Experiment zwischen der Schönen und dem Biest eine Interpolation zwischen zwei Extremen. Das Ergebnis ist eine überraschend neutrale kontrastarme Version.

Karloff explores the idea of irreconcilable differences, how two extremes could be combined into a coherent whole entity.

At the start the high-contrast Didot typefaces which are considered by many as some of the most beautiful in existence, and the eccentric Italian, a reversed-contrast typeface, which was designed to deliberately attract readers' attention by defying their expectations were looked at. No other style in the history of typography has provoked such negative reactions as the Italian.

Karloff, the result of this project, connects the high contrast modern type of Bodoni and Didot with the monstrous Italian. The difference between these attractive and the irritating forms lies in a single design parameter, the contrast between the fat and the thin. Having designed two diametrically opposite versions, a genetic experiment was undertaken with the offspring of the beauty and the beast, interpolation of the two extremes, which produced a surprisingly neutral low contrast version.

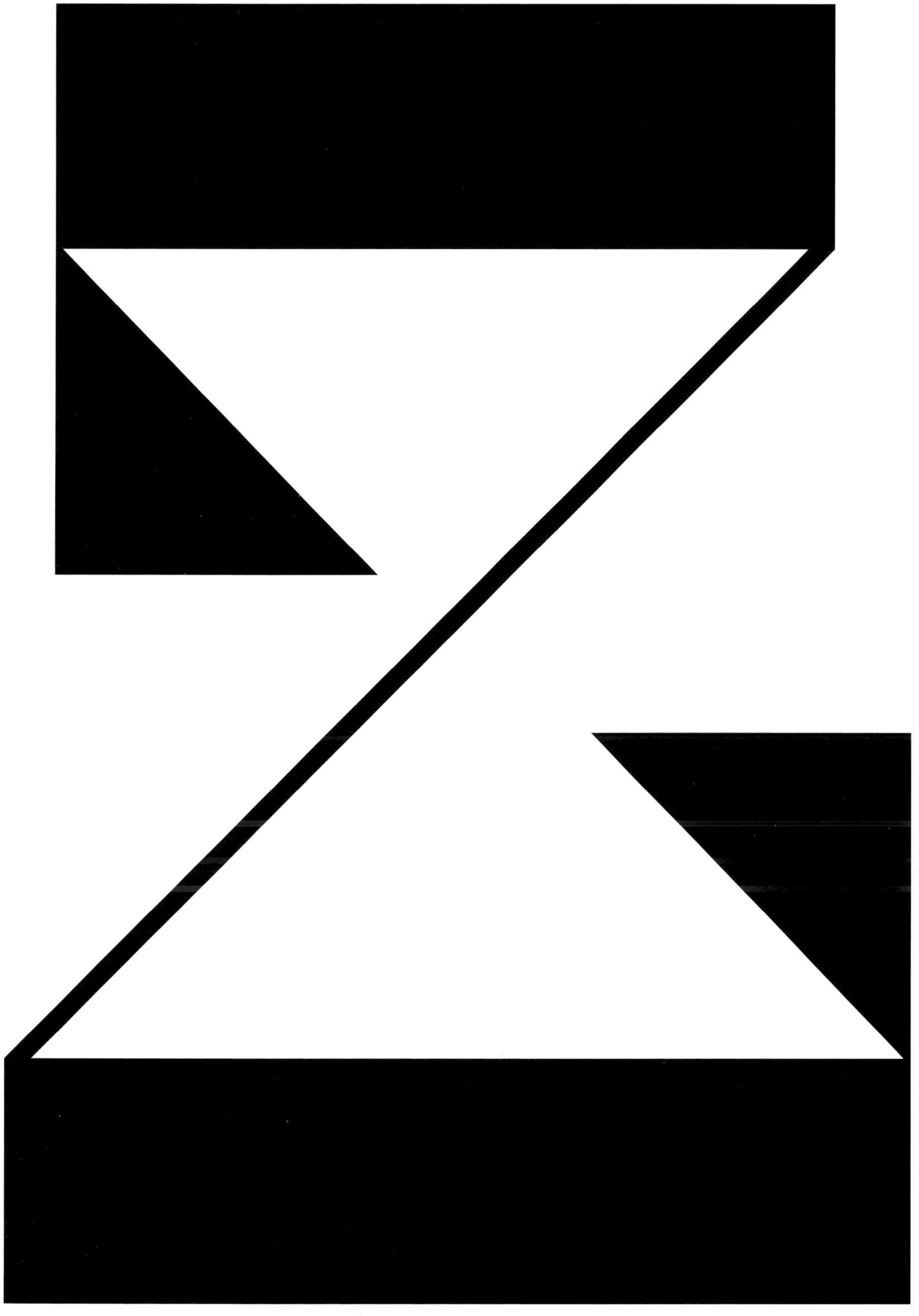

Der in der Tschechoslowakei geborene *Peter Bil’ak*[1] (*1973) studierte an der Jan van Eyck Akademie und ist heute als Grafiker und Schriftgestalter in den Niederlanden tätig. 1999 gründete er mit seiner Frau Johanna Bil’ak die Foundry »Typotheque«,[2] in der er seine Schriften, z.B. die Fedra,[3] Greta[4] oder Irma,[5] aber auch arabische Versionen seiner Schriften veröffentlicht. Bil’ak gilt als Pionier in der Entwicklung nicht-lateinischer Schriftsysteme. Als Mitgründer der »Indian Type Foundry« vertreibt er dort unter anderem seine Schrift Fedra Hindi. Bil’ak lehrt seit 2002 unter anderem an der KABK in Den Haag Typografie sowie Type & Media im Postgraduiertenkurs und hält Vorträge im In- und Ausland. Er ist Herausgeber zahlreicher Publikationen zum Thema Design, darunter »Dot Dot Dot«[6] und »Works That Work«.[7] Die Schrift *Karloff* entstand in Zusammenarbeit mit *Pieter van Rosmalen*.[8]

Woher kommt Ihr Interesse an Typografie?

pb: Ich lese und schreibe schon sehr lange, und habe mich immer für Text interessiert, und zwar aus zwei Gründen, nämlich erstens wegen des Aussehens, d.h. dem Schriftbild, aber auch wegen der Funktion von Text als Vermittler von Bedeutung und Information. Diese zwei sehr verschiedenen Aspekte können einem helfen, eine Arbeit sehr zu verbessern. Und meine Arbeit hatte gleich von Anfang an immer etwas mit diesen Aspekten zu tun, mit dem Schreiben und Entwerfen mit Texten. Das verstehe ich unter vollkommener Urheberschaft.

Sie interessieren sich für viele Dinge. Außer mit Lesen und Schreiben verbringen Sie auch viel Zeit mit anderen Interessen, wie Tanzen oder Kuratieren. Glauben Sie, dass Ihnen das hilft, ein guter Designer zu sein?

Ich weiß nicht, ob das jedermann hilft, aber für mich funktioniert es. Ich habe gemerkt, dass ich am zufriedensten bin, wenn ich etwas zum ersten Mal mache, und ich versuche, meine tägliche Arbeit danach zu gestalten. Um keinen zu gleichförmigen Job zu machen, wo ich immer wieder dasselbe mache. Ich wechsle meine Tätigkeiten gern ab. Manchmal kann das sehr spontan sein. Das ist interessanter, als nur das zu machen, was ich bereits kann, und dadurch

1 typotheque.com

2 Typotheque ist die Foundry von Peter Bil’ak, bei der er seine Schriften wie Fedra, Greta und Irma vertreibt. Auf der Website gibt es einen Blog, auf dem regelmäßig Neuigkeiten zu den Schriften vorgestellt werden.

3 Fedra ist eine Schrift, die zwei gegensätzliche Design-Ansätze bringt: die Steifigkeit einer Schriftart für den Computer-Bildschirm und die Flexibilität einer Handschrift. Typotheque, 2001, Peter Bil’ak.

4 Greta ist eine moderne Schriftfamilie, die speziell für die Anforderungen im Zeitungsdruck entwickelt wurde. Sie ist für den Einsatz bei kleinen Schriftgrößen optimiert. Typotheque, 2007, Peter Bil’ak.

5 Irma ist eine einfache und elegante Display-Schrift mit neun Schnitten. Die Schrift bietet verschiedene kontextuelle Buchstabenformen und erlaubt durch OpenType-Features die Schaffung von typografischen Mustern. Typotheque, 2009, Peter Bil’ak.

6 Dot Dot Dot ist eine Zeitschrift, die von Stuart Bailey und Peter Bil’ak produziert wurde, die zunächst Grafik-Design, später aber auch Themen aus den Bereichen Musik, Sprache, Literatur und Architektur behandelte. Nach zehn Jahren und 20 Ausgaben wurde die Zeitschrift 2011 eingestellt.

7 Das neue internationale Design-Magazin Works That Work wird von Peter Bil’ak herausgegeben und möchte Inspirationen und Beobachtungen kommunizieren — eine Art »National Geographic« für den Designbereich.

8 Pieter van Rosmalen betreibt das Design Studio CakeLab und arbeitet typischerweise an zehn verschiedenen Schriften gleichzeitig. Zu seinen Kunden gehören u.a. Firmen wie NBCUniversal, Audi AG und KPN. Er ist Mitbegründer der Type Foundry Bold Monday.

bekomme ich auch in diesen anderen Dingen Routine, und das ist mir wichtig. Ich will nicht nur eine Sache gründlich beherrschen, sondern ich möchte das breitere Spektrum meines Berufs kennenlernen.

Sie sind viel nach Indien gereist und haben dort eine Foundry aufgebaut. Sehen Sie sich oder Ihre Schriften als Botschafter von Sprache oder Kultur? Sind Sie oder Ihre Schriften Vertreter einer bestimmten Politik?

Nein, ich bin nicht politisch motiviert. Die Motivation ist eher, dass ich in Indien arbeite, einem Land mit endlosen Möglichkeiten. Dort leben eine Milliarde Menschen, und bisher ist auf dem Gebiet der Typografie noch nicht allzu viel passiert. Es ist wie eine Zeitreise, als wäre man im Europa des 17. Jahrhunderts. Und da wir inzwischen das 21. Jahrhundert haben, kann man für Indien viel Neues schaffen, was bisher aus verschiedenen Gründen nicht passiert ist.

Der Designprozess und die technischen Gegebenheiten sind komplizierter. Es gibt zehnmal so viele Schriftzeichen wie im Lateinischen, deshalb dauert es viel länger, eine neue Schrift zu entwickeln. Aber es gibt noch keinen Markt dafür, also investieren die Firmen auch nichts in Typografie. Aber es lohnt sich, dran zu bleiben, denn es ist nur eine Frage der Zeit, bis auch hier Typografie wichtiger wird. Ein weiterer Grund, nicht aufzugeben, ist, dass viele Sprachen aussterben, weil sie nicht in schriftlicher Form existieren. Außerdem gibt es nicht genug Fonts. Hier hat man die Möglichkeit, wirklich etwas zu bewirken, und nicht nur weitere Schriften zu schaffen.

Wir haben auch Nadine Chahine[9] interviewt, die für ihre arabischen Schriften bekannt ist. Sie kennt beides, die lateinischen und die arabischen Schriften. Sie schufen den Font Fedra Hindi. Muss man eigentlich eine Sprache und Schrift kennen, um ihre Schriftzeichen zu verändern? Wie haben Sie das gemacht, wo fangen Sie an?

Schreiben und Lesen sind zwei sehr unterschiedliche Vorgänge. Schließlich kann man eine Sprache beherrschen, ohne sie zu schreiben, und umgekehrt. Einer der besten Schriftdesigner, die es je gab, war Edward Pradell, ein spanischer Drucker, der die schönsten Schriften entwickelte. Und das, obwohl er ein Analphabet war. Also ist es möglich, mit Schriftzeichen zu arbeiten, ohne ihre Bedeutung zu verstehen. Es ist nichts weiter als ein Spielen mit schwarzen und weißen Formen.

Das einzig Wichtige ist die Struktur, und zu wissen, wie die Sprache formell funktioniert. Ich habe ebenfalls arabische Fonts geschaffen, also weiß ich, was Nadines Arbeit bedeutet. Bei dieser Arbeit muss man sich der Möglichkeiten einer Schrift stärker bewusst sein. Man muss herausfinden, was schon gemacht worden ist und was nicht. Was waren Fehlschläge, was war erfolgreich? Es ist viel besser, mit einem leeren Blatt Papier anzufangen und keine vorgefassten Meinungen zu haben. Nadine ist Libanesin. Und sie sieht die Dinge aus der Warte einer Libanesin. Ich glaube, es ist schwer, sich davon zu lösen. Wenn man mit jemandem aus dem Iran spricht, würde der vielleicht sagen, dass diese Sichtweise falsch ist, weil man da wieder eine ganz andere Sicht der Dinge hat. Deshalb hilft es, wenn man von außen kommt. Man sieht die Dinge auf andere Art und Weise und findet seinen eigenen Weg. Aber nichtsdestoweniger arbeite ich mit vielen Menschen zusammen, welche die Sprache sprechen und mit denen ich über die Details diskutieren kann. Ich spreche viel mit Menschen, auf diese Weise kann ich deren Kenntnis der Sprache mit meiner Kenntnis der Typografie und Technologie verbinden.

Aber wie ließ es sich »übersetzen«? Inwiefern ähneln sich Fedra und Fedra Hindi?

Für Fedra Hindi habe ich mir keine digitalen Fonts angesehen. Ich fing damit an, dass ich mir Handschriften ansah, weil sie noch keine Spuren technologisch bedingter Veränderungen aufweisen. Bei digitalen Fonts läuft man Gefahr, den falschen Weg einzuschlagen, weil man sich nach dem richtet, was in der Vergangenheit gemacht wurde. Einerseits gibt es die gesprochene Sprache, die sich über tausende von Jahren nicht verändert hat, andererseits gibt es die geschriebene Form, die sich im Laufe der Zeit verändert hat. Ich sammelte also viele verschiedene

9 Nadine Chahine (→ S. 80)

Handschriften von Leuten aus der Region. Dann verglich ich sie und sah mir die verschiedenen Formen an.

Schließlich druckte ich eine große Anzahl davon aus und entwickelte sie weiter. Und basierend auf diesen Zeichnungen bemühte ich mich um eine vereinfachte Version, zu der die Design-Parameter von Fedra passten.

Es gibt zwei Komponenten. Im Grunde genommen spielt man mit Schriftzügen. Es ist wichtig, dass man keine Schlüsse aus der eigenen Sprache zieht, denn hier hat man es mit etwas zu tun, was völlig außerhalb des gewohnten Einflussbereichs liegt. Ich will keine fremden Elemente hineinbringen, ich will von etwas anderem ausgehen und eine natürliche Ausdrucksform finden. Meist entscheide ich mich für eine Multiscript-Typografie, bei der ich verschiedene Schriften kombiniere. Das ist sehr eigenwillig, ohne dass sie dann alle gleich aussehen. Es geht nicht darum, Formen zu kopieren, das ist gar nicht so wichtig. Es geht mehr darum, sie zu verstehen und so zu formen, dass ihre Bedeutung nicht verlorengeht.

Ihre Arbeit hat Pioniercharakter. Wie wird sich die Typografie in Indien oder den arabischen Ländern im Laufe der nächsten Jahre entwickeln? Wird Ihre Arbeit die Entwicklung beeinflussen, wird sie etwas verändern?

Jedes Mal, wenn ich in den letzten Jahren in Indien war, merkte ich, dass sich riesige Veränderungen vollzogen hatten. Als ich das erste Mal dort hinkam, hatte ich es noch schwer, Schriftdesign zu erklären, jetzt gibt es überall Schriftdesigner. Das ist innerhalb von fünf Jahren passiert. Anfangs war die Diskussion über das, was wir machen, eine völlig andere. Damals gab es einfach nicht genug Schriftdesigner. Es war ein gewaltiger Prozess. Dasselbe passiert in Nahost mit den arabischen Schriften. Die Menschen fangen an, sich dafür zu interessieren, weil es sich verbreitet und neue Möglichkeiten schafft. Natürlich macht es Spaß, daran beteiligt zu sein, denn solche Anfänge sind immer spannend. Es gibt so vieles, was noch nicht ausgelotet ist. Diese Entwicklung wird sich fortsetzen, und sie wird nicht nur von einheimischen Designern, sondern auch von Außenstehenden beeinflusst werden. Es ist gut, dass beide daran beteiligt sind, und es wird sich vieles verändern, egal, ob ich dazu beitrage oder nicht.

Die Schrift, die Sie zeigen, ist Karloff. Können Sie uns etwas über dieses Projekt erzählen?

Die Geschichte von Karloff fing mit einem Vortrag an, den ich in Kopenhagen bei der Konferenz »Conceptional Typography« hielt. Ich sprach darüber, dass Schriftdesign immer noch ein Handwerk ist, welches besondere Fähigkeiten voraussetzt und dass das Übertragen einer Idee in eine funktionierende Schrift in dieser Disziplin weiterhin ein kritischer Punkt ist. Ideen allein reichen nicht für Schriftdesign. Man kann keine Vorstellung von etwas haben, ohne die Form festzuhalten. Man muss imstande sein, sich visuell zu artikulieren. Ich habe also bei dieser Konferenz behauptet, dass es so etwas wie konzeptionelle Typografie nicht gibt. Ich fing an, mir nicht nur die Formen einer Schrift anzusehen, sondern ich wollte die Beziehungen der Formen zueinander verstehen. Nicht lange, bevor ich Karloff entwickelte, schuf ich Greta, eine Schrift in vielen verschiedenen Gewichten und Stilen. Das Entwickeln einer Schrift ist ein Designprozess, bei dem es um die Beziehungen der verschiedenen Stile untereinander geht.

Bei Karloff war es derselbe Gedanke, nur noch drastischer. Bei Greta bestehen die Verwandtschaften durch Breite und Gewicht, damit arbeitet man am häufigsten im Schriftdesign. Karloff hat zwei Designparameter. Etwas, das gleich ist, und etwas, das ganz anders ist. Ich wollte am Ende zwei verschiedene Schriftformen haben. Ich war fasziniert von der Idee, zwei diametral unterschiedliche Schriften miteinander zu kombinieren. Das erwies sich als schwierig, also entschied ich mich für eine Schrift, die nicht zu gegensätzlich war. Mein Projekt fing damit an, dass ich mir Schriftbeispiele ansah, und zwar Beispiele aus der Geschichte der Typografie, die als besonders schön oder besonders hässlich galten. Im 18. Jahrhundert waren Bodoni[10] und Didot[11] die schönsten Schriften, die es gab. Man schätzte sie vor allem wegen ihrer perfekten Handwerkstechnik, denn inzwischen war die Technologie soweit vorangeschritten, dass Dinge möglich waren, die man vorher nicht machen konnte.

Während der industriellen Revolution waren ein paar Negativschriften entwickelt worden. Damals hatte man es schwer, auf sich aufmerksam zu machen. Also fing man an, die Schriften schwerer zu machen, extra bold, oder ganz schmal, oder unterstrichen – man versuchte alles, um Aufmerksamkeit zu bekommen und die Welt mit neuen Schriften zu überraschen. Diese Strategie funktionierte auch eine kurze Zeit. Konventionelle Erwartungen wurden herausgefordert, indem man Schwarz und Weiß vertauschte. Was bisher dick war, wurde dünn. Dies waren für mich die hässlichen Beispiele.

Wenn man beide Fonts vergleicht und sich die zugrundeliegende Struktur ansieht, sind sie identisch. Es gibt Parameter, die sind gleich geblieben, bis auf einen. Der Unterschied zwischen den schönen und den hässlichen Formen liegt in einem einzigen Designparameter, nämlich dem Kontrast zwischen dick und dünn. Und das ist im Grunde eine Umkehrung des Prozesses, einen Kontrast herzustellen.

Was würden Sie in Karloff setzen? Für welche Art von Projekten ist sie gedacht? Sind Sie manchmal enttäuscht, dass sie nicht so angewandt wird, wie Sie es sich vorgestellt haben?

Eigentlich nicht. Ich habe Glück, weil viele gute Designer unsere Schriften kaufen und die verwenden sie intelligent. Manchmal gebe ich diese Fonts Freunden, die exzellente Designer sind. Es gibt den Kunden Anregungen, was man mit Schriften alles machen kann, und auf diese Art und Weise kann ich etwas Einfluss nehmen. Ich glaube, Karloff eignet sich gut für redaktionelle Texte, besonders wenn es darum geht, zwei verschiedene Stimmen deutlich zu machen. Aber auch hier hängt es letztendlich vom Designer ab, was er damit macht.

Das Interview mit Peter Bil'ak führten Anna Alexander, Lisa Grünwald und Bahar Hasan.

10 Giambattista Bodoni (1740–1813) hat eine Reihe klassizistischer Antiquaschriftarten geschaffen, die als Bodoni bezeichnet werden. Sie verfügen über einen hohen Strichstärken-Kontrast und flache Serifen (→ S. 36).

11 Didot. Die klaren Formen dieses Alphabets verdeutlichen objektive und rationale Eigenschaften und reflektieren die Ideale der Aufklärung.

Peter Bil'ak[1] (*1973), who was born in Czechoslovakia, studied at the Jan van Eyck Academy. Currently he is working as graphic and type designer in the Netherlands. In 1999, he and his wife, Johanna Bil'ak, established the »Typotheque«[2] foundry. There he publishes his fonts, e.g. Fedra,[3] Greta,[4] and Irma,[5] and Arabic versions of his typefaces. Bil'ak is considered a pioneer in the development of non-Latin fonts. As co-founder of the »Indian Type Foundry« he sells, amongst others, his Fedra Hindi font. Since 2002, Bil'ak has been teaching at KABK in Den Haag Typography, Type & Media for a post-graduate course and other institutions. He has given lectures in the Netherlands and abroad and is the editor of numerous publications on the topic of design, like, for example, »Dot Dot Dot«[6] and »Works That Work«.[7] *Karloff* was developed together with *Pieter van Rosmalen*.[8]

Where does your interest in typography come from?

pb: I have been reading and writing for a long time and have always been interested in text from both points of view. In the appearance of a text, i.e. the typeface; but also in text as a carrier of meaning and information. These two different aspects can help to improve your work. Right from the start my earliest projects always had something to do with those aspects, writing and designing with text. This is what I understand by complete authorship.

You are interested in a lot of things. Besides reading and writing, you spend quite a lot of time doing other things, like dancing or curation work. Do you think this helps you to be a good designer?

I don't know if it helps for everyone, it works for me. I realized that I am usually happiest when I do things for the first time. I try to carry this over into my daily work. Not to have too routined a job, not to repeat the same thing over and over. I keep changing disciplines. Sometimes it is very spontaneous. That is more interesting than doing things I know and it makes me an expert of those things, which I think is important. Not just to know a thing in depth, but also to appreciate the wide spectrum of a profession.

You've been travelling a lot to India and established a foundry there. Do you see yourself/your typefaces as an ambassador of language or culture? Are you or your typefaces a political representative?

No, there is no political motivation. It's the motivation of being in India, a country with enormous possibilities. There are one billion people, and so far not a lot has happened about typography. It's like a journey through time, like being in the Europe of the 17th century. Having experienced the 21st century, you are able to do a lot of new things for India which for many reasons haven't been done so far.

The design process and the technical aspects are more complicated. The character set is ten times as big as the Latin one. Therefore it requires much more time to create a typeface. Also there is no market, which is why big companies don't invest in typography. But it is worth trying because it is only a question of time before typography becomes more important. Another reason to persevere is that a lot of languages are dying out because they do not exist in a written form. Also there are not enough fonts. You have the opportunity to make a real difference in design, rather than just creating yet another typeface.

We also had an interview with Nadine Chahine,[9] who is famous for her Arabic typefaces. She knows both scripts, Arabic and Latin. You made the font Fedra Hindi. Is it necessary to know language and script in order to »transform« the letters? How did you do that? What do you go for?

Writing and reading are separate things. After all, you can speak a language without being able to write it and vice versa. One of the best type designers in history was Edward Pradell, a Spanish printer who produced one of the most beautiful typefaces. And yet he was illiterate, he couldn't read at all. So it is possible to work with letterforms without understanding their meaning. In the end it's a formal play of black and white.

The only important thing to know is the structure and how language works formally. I have been designing Arabic

1 typotheque.com

2 Typotheque is Peter Bil'ak's foundry, selling his typefaces Fedra, Greta and Irma. His website includes a blog where he publishes regular news concerning his typefaces.

3 Fedra is a typeface combining two contrasting design principles: the stiffness of a type suitable for the screen and the flexibility of script. Typotheque, 2001, Peter Bil'ak.

4 Greta is a family with a modern typeface which was developed especially for the requirements of newsprint. It has been optimized for small-size print. Typotheque, 2007, Peter Bil'ak.

5 Irma is a simple and elegant display typeface with nine styles. It offers several contextual letter forms and, through OpenType features, allows the creation of typographic patterns. Typotheque, 2009, Peter Bil'ak.

6 Dot Dot Dot was a magazine produced by Stuart Bailey and Peter Bil'ak and addressing subjects from graphic design, later also from the areas of music, language, literature and architecture. After ten years and 20 issues the magazine was discontinued in 2011.

7 The new international design magazine Works That Work is edited by Peter Bil'ak and wants to communicate inspiration and observations – a kind of »National Geographic« of Design.

8 Pieter van Rosmalen runs a graphic design studio called CakeLab and is typically working on ten different typefaces at the same time. He has worked on custom typefaces for worldwide clients such as NBCUniversal, Audi AG and KPN. He is a partner at the type foundry Bold Monday.

9 Nadine Chahine (→ S. 85)

The brain you stole, Fritz.
Think of it.
The brain of a dead man waiting
to live again in a body
I made with my own hands!

Karloff Positive Regular, Bold – 13pt

Look! It's moving.
It's alive. It's alive.
It's alive, it's moving,
IT'S ALIVE!

Karloff Neutral Std. Regular, Italic – 18pt

The neck's broken. The brain is useless. We must find another brain.

Karloff Negativ Bold – 30pt

Crazy, am I?
We'll see whether
I'm crazy or not.

Karloff Negative, Neutral, Positive Regular – 50pt

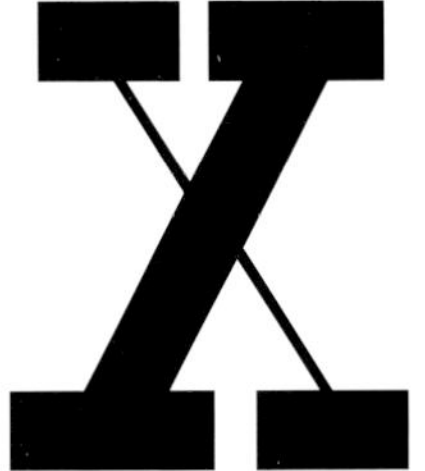

Karloff Negative

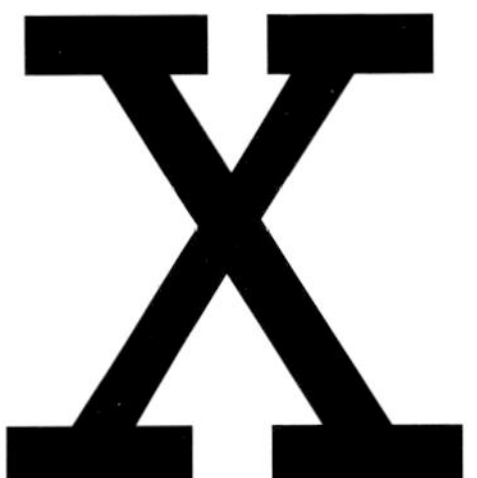

Karloff Neutral

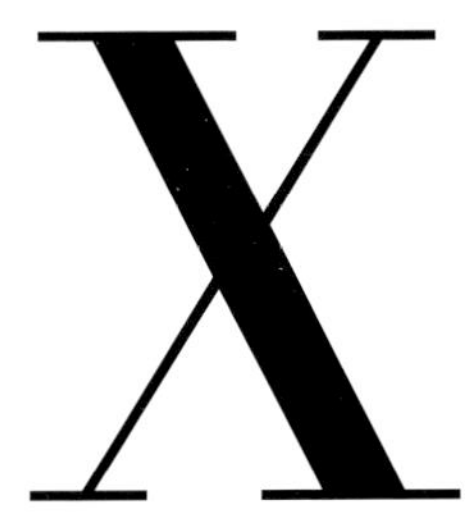

Karloff Positive

»In the end, it's a formal play of black and white.«

Peter Bil'ak
→ S. 136

fonts too, so I know what Nadine's work is about. In that process you have to be more concerned about the possibilities of the script. You need to find out what has been done and what hasn't. What were the failures and what were the successes? It is much better to start with a blank sheet of paper because you don't have preconceived ideas about it. Nadine is Lebanese. She has this Lebanese way of looking at things. I think it is very hard to leave this behind. If you talk to an Iranian person, they would probably say this way of looking is wrong because they would have a completely contradictory view on things. Being on the outside helps. You have a different way of seeing things and finding your own way. Having said this, I work with many different people who speak the language, and with whom I can discuss details. I always talk to people, in order to try and combine their knowledge of the language with type and technology.

But still, how did it »translate?« What are the similarities between Fedra and Fedra Hindi?

For Fedra Hindi I didn't look at digital fonts. I first started to look at handwritings because they don't have the traces of technological limitations. Digital fonts tend to lead you the wrong way because of what had been done in the past. On the one hand there is the spoken language which hasn't changed in thousands of years, but on the other hand there is a written form which has slowly been adapted. I collected different types of handwriting from a lot of people in different regions. Comparing them, I looked at the different shapes. Finally I printed a large selection and refined them. Based on those drawings I tried to find a simplified version, which would work with the design parameters of Fedra.

There are two components. It's basically a play of strokes. It's important not to make assumptions based on your own language or knowledge, because you deal with something which is outside your own sphere. I don't want to bring foreign elements into it, I want to go from inside out and find a natural way of expressing the language. Creating multiscript typefaces where one script imposes formal attributes on another script is an ill-considered way. It's not about copying shapes. The forms are not very important. It's more about understanding the shapes and forming them so that the intention isn't lost.

Your work has pioneer character. How will typography in India or Arab countries develop over the next few years? Will your work have any influence, will it change anything?

Going back in India over the last few years, I realized that each time there were enormous changes. The first time I came it was hard to explain type design, but now there are type designers everywhere. That change happened within five years. At the beginning there was a totally different discussion about the work we do. There just weren't enough designers back then. It has been a big process. The same happens in the Middle East with the Arabic typefaces. People start getting interested because it propagates and creates new opportunities. Of course it's fun to be part of it because the early stages are the most exciting ones. There are many things that haven't yet been explored. This will continue, and it will be influenced by outsiders, but also by local people. It is good to see both, but it will change, whether I am involved or not.

The typeface you show is Karloff. Please, could you tell us something about that project?

The story about Karloff started with a lecture I gave at the »Conceptional Typography Conference« in Copenhagen. I talked about how type design still is a craft based on discipline which requires a particular set of skills, so the process that transforms the pure idea into a functional font is a critical part of the discipline. Ideas alone don't work for type design. You cannot just have a concept without capturing its form. You need to be able to propose and articulate visually. So in that conference I argued that there is no such thing as conceptual type. And I said that it would be very interesting to see an example of this. I became interested in not just looking at shapes of type but understanding the relationships between forms. A short time before Karloff I made Greta, which is a typeface in many different weights and styles. Designing a typeface is about designing the relationship between different styles. With Karloff it was the same idea, only more extreme. Greta is related through width and weight, the most conventional way of working with type design. Karloff has two design parameters. There is something shared and something different. I wanted to end up with two different forms of type. I was fascinated by the idea of formally connecting two completely different typefaces, the opposites of each other. That turned out to be difficult, so I took something not too contradictory. The project started by looking at the most beautiful examples of type, and for that I was trying to find examples of beauty and ugliness in history. In the 18th century Bodoni[10] and Didot[11] were the most beautiful typefaces in existence. They were mainly appreciated for their level of craftsmanship. By now the technology was advanced enough that it was possible to do something which had been impossible before.

During the industrial revolution some new reversed-contrast typefaces had been designed. In those days it was very hard to get attention. People tried to make typefaces heavier, extra bold, condensed or underlined, trying anything to get attention and surprise the world with new and different forms. It was a technique which worked quite well in the short term. Conventional expectations were challenged by reversing constructional type. What was thick became thin. For me, that was the ugly aspect of it.

If you compare both fonts and look at the underlying structure, they are identical. There are parameters which stayed the same, except for one thing.

The difference between the attractive and the ugly forms lies in a single design parameter, the contrast between thick and thin. And this is basically reversing the process of creating a contrast.

What would you set in Karloff? What kinds of project is it for? Are you sometimes disappointed that it is not being used as you intended?

Not really. I'm very lucky that many very good designers buy our fonts and use them quite well. Sometimes I give these fonts to my friends, who are often excellent designers. It gives other users certain suggestions of what you can do with typefaces. In this way you can exercise a little influence. I think Karloff can be very good for editorial use, especially when you have to express two different voices. But again, in the end it depends on the designer what he does with it.

Peter Bil'ak was interviewed by Anna Alexander, Lisa Grünwald and Bahar Hasan.

10 Giambattista Bodoni (1740–1813) has created a number of classicistic Antiqua types, known as Bodoni. They have a high contrast in stroke widths and flat serifs (→ S. 39).

11 Didot. The clear forms of this alphabet illustrates objective and rational qualities and reflects the ideals of the Age of Enlightenment.

KARL NAWROT & RADIM PEŠKO

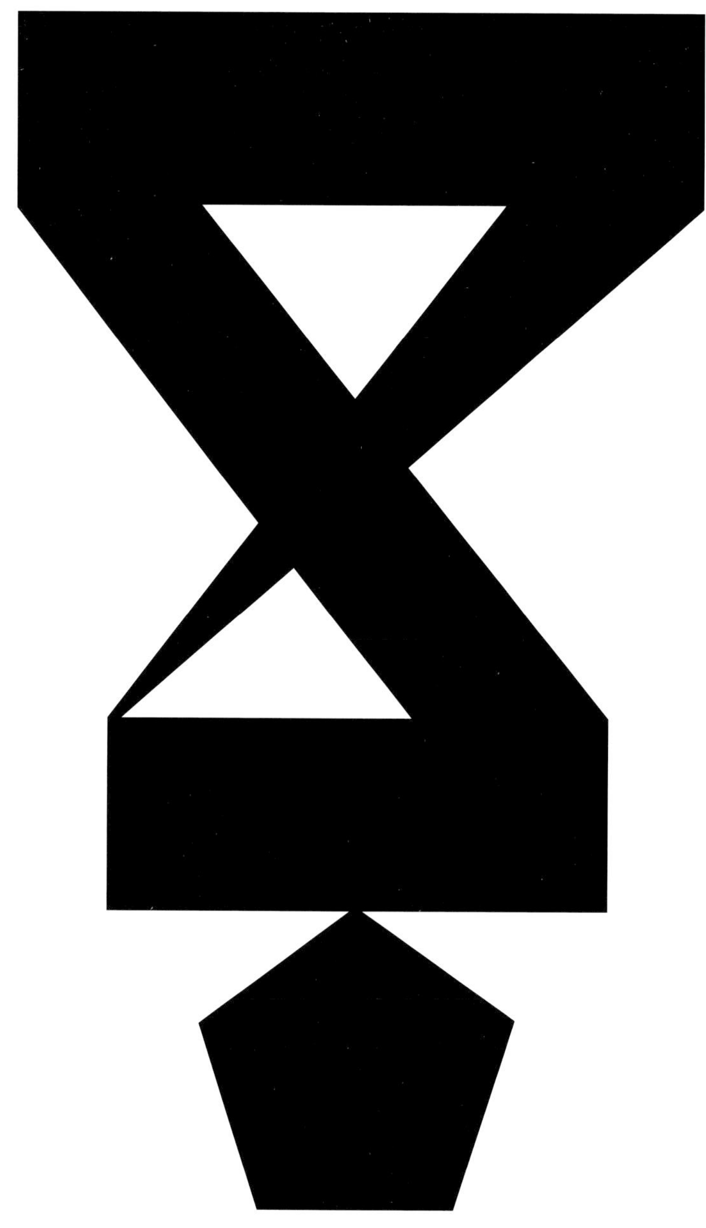

Schrift. Typeface.	Lÿno
Gestalter. Designer.	Karl Nawrot & Radim Peško
Label. Foundry.	RP Digital Type Foundry radimpesko.com
Jahr. Year.	2010–2012

Lÿno wurde zwischen 2009 und 2012 von Karl Nawrot und Radim Peško gestaltet. Die Schrift liegt in vier Schnitten vor. Die Schriftzeichen von Ulys(ses 31), Stan(ley Kubrick), Jean (Arp) and Walt (Disney) verbinden die digitale Freiheit der Produktion mit verspielten Typologien und Formen. Ihre Formen sind offen und vielfältig und ihre Absicht ist wie folgt: Widerstand gegen alle normativen Tendenzen und Verneinung der definierten Form.

Lÿno was designed by Karl Nawrot and Radim Peško between 2009 and 2012. It is available in four styles. The characters of Ulys(ses 31), Stan(ley Kubrick), Jean (Arp) and Walt (Disney) negotiate the digital freedom of their production through playful typologies of form. They are open and various, and their spirit is this: to resist normative tendencies and to reject the idea of definitive form.

WALT, JEAN, ULYS & STAN

ABCDEFGHIJKLMN
OPQRSTUVWXYZ

Lÿno Walt

ABCDEFGHIJKLMN
OPQRSTUVWXYZ

Lÿno Jean

ABCDEFGHIJKLMN
OPQRSTUVWXYZ

Lÿno Ulys

ABCDEFGHIJKLMN
OPQRSTUVWXYZ

Lÿno Stan

Radim Peško[1] (*1976 in Kyjov, Tschechoslowakei) arbeitet als Grafik-Designer in Amsterdam. Nach seinem Studium an der Academy of Arts in Prag und der Tätigkeit bei »LCP« in London, beendete er 2004 sein Master-Studium am Werkplaats Typografie in Arnheim. Er arbeitet als Type Designer sowie an zahlreichen Editorial- und Ausstellungsprojekten. 2010 gründete Peško seine eigene Foundry »RP-Type«, die sich auf Schriften konzentriert, die sich durch formale wie auch konzeptuelle Stringenz auszeichnen. Peško lehrte an der Rietveld Academie in Amsterdam und an der École Nationale Supérieure des Beaux-Arts de Lyon.

Während unserer Nachforschungen hatten wir den Eindruck, dass Sie ein wenig geheimnisvoll taten. Gibt es dafür einen besonderen Grund?

rp: *(Lächelt.)*

Wie werben Sie für Ihre Font-Designs und für Ihren Schriftvertrieb?

Gar nicht, es passiert von allein durch meine Arbeit oder verwandte Projekte.

Wird Ihr Design durch die neuen Technologien (neue Software, neue Möglichkeiten durch OpenType-Programmierung[2] usw.) beeinflusst? Und wenn ja, wie?

Nun, das weiß ich nicht so genau, vermutlich ist es wie bei allen anderen auch, die in ihrer Arbeit davon abhängig sind. Ich glaube nicht, dass sie mein Denken oder meine Ideen beeinflussen, meist sind sie einfach nur hilfreiche und freundliche Gefährten im Arbeitsprozess.

Interessieren Sie sich für eine Optimierung der digitalen Medien oder der Web Font-Technologie? Spielen diese Themen eine Rolle für Sie? Und warum oder warum nicht?

Natürlich, aber das ist ein so großes Thema voll technischer Einzelheiten, dass es in diesem Zusammenhang weniger von Interesse ist. Ganz allgemein glaube ich, dass dieser Aspekt in Zukunft immer weniger ein Thema sein wird, weil wir selbst immer vertrauter damit umgehen.

Ist der Computer Freund oder Feind des Schriftdesigners?

Der Computer ist ein Werkzeug. Ob Freund oder Feind hängt davon ab, wie man damit umgeht. Sagt man das nicht auch vom Alkohol?

Wie beeinflusst Ihre Arbeit, die Sie neben dem Schriftdesign noch machen, Ihr Schriftzeichnen?

Ich habe fast sieben Jahre an der Gerrit Rietveld Academie unterrichtet. Jetzt habe ich beschlossen, mal eine Pause zu machen und mich mehr auf die Praxis zu konzentrieren. Ich fotografiere genauso gern wie ich zeichne, und irgendwie kommt beim Schriftdesign ja auch

1 radimpesko.com

2 OpenType ist ein aktuelles Datenformat für digitalisierte Satzschriften, von Adobe und Microsoft entwickelt, welches die Grenzen der weit verbreiteten Font-Formate PostScript (Type 1) und TrueType sowie den limitierten Zeichenvorrat von PC/MAC überwindet (→ S. 161).

»THE COMPUTER IS A TOOL. IF IT IS FRIEND OR ENEMY DEPENDS PROBABLY ON THE PERSON WHO HANDLES IT. BUT ISN'T THAT WHAT THEY SAY ABOUT ALCOHOL, TOO?«

LŸNO

Radim Peško
→ S. 146

beides zusammen, nur kann man mit Fotografien Geschichten ganz anders erzählen. Schließlich macht das Schriftdesign ja nur ein Drittel meiner Arbeit aus, daneben mache ich noch Auftragsarbeiten, ich arbeite zusammen mit anderen Künstlern oder helfe bei der Organisation der Graphic Design-Biennale[3] in Brünn / Tschechien.

Was würden Sie dann als Ihren eigentlichen Beruf bezeichnen? Sind Sie in erster Linie Schriftdesigner / Fotograf / Lehrer / Künstler? Ist Schriftdesign für Sie eine Art Ausgleich zur Auftragsarbeit, oder ist es eher umgekehrt?

Ich glaube, da sehe ich die Dinge etwas pragmatischer: Es geht doch immer darum, innerhalb einer bestimmten Zeit, für eine bestimmte Situation oder zu einem bestimmten Zweck das Bestmögliche zu schaffen, deshalb nenne ich alles »Arbeit«. Ob meine Fähigkeiten als Schriftdesigner meine Fähigkeiten als Fotograf übertreffen? Das ist natürlich eine andere Frage …

Was ist der Unterschied zwischen aktuellem Grafik-Design und den aktuellen Schriften? Wie beeinflussen sich Ihrer Meinung nach diese beiden Gebiete? Ergeben sich die Trends im aktuellen Grafik-Design aus den Trends im aktuellen Schriftdesign, oder ist es umgekehrt? Wäre das eine ohne das andere möglich?

Ich würde sagen, ein Schriftentwurf kommt auch ohne Grafik-Design aus; aber Grafik-Design nicht ohne Typografie.

Es gab mal eine Zeit, als Schriften nur entwickelt wurden, wenn man sie für einen besonderen Zweck brauchte. Heutzutage entwickeln viele Schriftdesigner z.B. klassische Schriftbilder oder Display-Schriften, obwohl es dafür schon eine unendliche Anzahl von Schriften gibt. Was ist die Motivation dahinter?

Das ist eine komische Frage, denn das klingt ja, als habe bei der Erschaffung der Welt jemand festgelegt, wie viele Schriften es geben dürfe, und wenn diese Anzahl erreicht sei, gebe es keinen Grund, noch weitere zu schaffen. So wie sich die Sprache fortwährend verändert (z.B. im Zusammenhang mit dem technologischen Fortschritt), so werden auch neue Schriften gebraucht.

Es werden online, z.B. unter myfonts.com,[4] eine unglaubliche Anzahl neuer Schriften angeboten. Die meisten gibt es in großen Paketen mit vielen Schnitten. Ihre Schriften dagegen gibt es meist nur mit einer kleinen Auswahl an Schnitten, oft sogar nur mit einem.

Ich bin eben sehr langsam.

Wie viele Schnitte braucht eine Schrift? Sind das nur Verkaufsstrategien oder sind die wirklich notwendig? Brauchen wir diese »Super-Familie«, oder ist irgendwo eine Grenze?

Es kommt darauf an … und vielleicht gibt es wirklich keine Formel dafür. Ich persönlich glaube nicht, dass die Pakete, die Sie »Super-Familien« nennen, wirklich das beinhalten, was sie versprechen. Manchmal glaube ich wirklich, dass sie allein aus dem Grund existieren, weil es technisch möglich ist, sie zu schaffen.

Haben Sie eine besondere Zielgruppe? Wer sollte Ihrer Meinung nach Ihre Fonts benutzen, und für welche Zwecke?

Ich denke nicht an einen »Markt« oder an »Zielgruppen« – ich glaube, das wäre für mich das Ende als Schriftdesigner. Ich betrachte Schriften als selbstständige Arbeiten. Entweder die Leute finden selbst heraus, wie sie sie anwenden können, oder sie tun es nicht. Arbeiten, die ich vor acht Jahren gemacht habe, wurden erst kürzlich entdeckt und benutzt. Man muss Geduld haben.

3 Die Internationale Biennale für Grafik-Design in Brno/Brünn (CZ), 1963 gegründet, ist eine der ältesten Veranstaltungen im Bereich des Grafik-Designs weltweit.

4 MyFonts ist eine digitale Schriftenplattform mit Sitz in Woburn, Massachusetts, welche Schriften über die Internetseite myfonts.com verkauft. Sie wurde von Bitstream Inc. im September 1999 während der ATypI Konferenz in Boston initiiert und verkauft seit März 2000 Schriften.

In den letzten Jahren sind viele neue Schriftvertriebe auf dem Markt erschienen, besonders in der Schweiz und in den Niederlanden. Was ist der Grund dafür? Ist das nur ein Trend oder eine längerfristige Entwicklung?

Ich denke, es hängt mit der Technologie, der Erreichbarkeit und der Wirtschaftlichkeit zusammen. Dasselbe passiert ja auch auf anderen Gebieten – zum Beispiel in der Musik, beim Film und im Verlagswesen. Ich sehe nicht, warum ausgerechnet Schriftdesign da eine Ausnahme machen sollte.

Zweitens sind ja die Schweiz und die Niederlande schon seit den Dreißigerjahren führend im europäischen Design, also überrascht es mich nicht, dass viele neue Schriftvertriebe von dort kommen. Andererseits werden unsere Grenzen zunehmend durchlässiger und so kann es passieren, dass auch Leute aus anderen Ländern etwas dazu beitragen, während die traditionellen Leitfiguren irgendwann anfangen, sich zu wiederholen oder müde werden. Wer weiß …

Sie leben im Moment noch in Amsterdam. Welche Wirkung hat diese Stadt auf Ihre Arbeit? Ist es wichtig für einen jungen Schriftdesigner, in Amsterdam zu leben, oder in Lausanne, oder Den Haag …?

Natürlich hat die Stadt, in der man lebt, einen großen Einfluss auf das eigene Leben. Amsterdam ist eine Stadt, die inspiriert – aber nicht unbedingt vom Standpunkt des Schriftdesigners, denn hier gibt es keine große Schrift-Szene (vielleicht ist das genau das, was mich an Amsterdam so reizte).

Aber ich werde ja wegziehen. Wohin, ist natürlich noch ein Geheimnis …

Das Interview mit Radim Peško führten David Dusanek und Matthias Dufner.

Radim Peško[1] (*1976 in Kyjov, Czechoslovakia) is a graphic designer based in Amsterdam. After his studies at the Academy of Arts in Prague and »LCP« London he completed his Master degree at Werkplaats Typografie in Arnhem in 2004. He works in the field of type design, editorial and exhibition projects. In 2010 he established his own Digital Type Foundry »RP-Type« that specializes on typefaces that are both formally and conceptually distinctive. He teaches at the Rietveld Academie in Amsterdam and the École Nationale Supérieure des Beaux-Arts de Lyon.

During our research we had a feeling as if you presented yourself a little secretive. Is there a special reason for that?

rp: *(Smile.)*

How do you advertise your font designs and your foundry?

I don't, it happens just through my work or related projects.

Do the new technologies (new software, new features through OpenType[2] programming, etc.) affect your designs? And if so, how?

Well, I don't know, I guess it's like with anybody else who depends on them for their work. I don't think they influence my thinking or initial ideas, they are just rather helpful and friendly (mostly) companions in the work process.

Are you interested in optimization for digital media or web font technology? Do these topics play a role for you? Why or why not?

Naturally, but that is a big subject full of technical details which I don't think are so interesting in this context. In general, I think we will speak less and less about this technology aspect in the future, because we ourselves will be much more familiar with it.

Is the computer the type designer's friend or enemy?

The computer is a tool. If it is a friend or an enemy depends on the person who handles it. Isn't that what they say about alcohol, too?

How does your work, besides type design, influence your font sketches?

I used to teach at the Gerrit Rietveld Academie for almost seven years. Now I've decided to quit regular teaching for a while, move on and focus more on my practice. I like picture making as much as drawing, and in a way they both come together in type design, but with photographs you can tell stories in a different way. After all, type design occupies only one third of my practice, besides that, I am doing commissioned work, collaborate with fellow artists or co-organise the Graphic Design Biennial[3] in Brno, Czech Republic.

So what would you say is your regular occupation? Are you above all a type designer/a photographer/a teacher/an artist? Is type design some kind of counter-balance to your commissioned work or is it the other way round?

I am afraid I see things a bit more pragmatic: It's all about doing something valuable at a given time, a given situation or for a certain purpose, so I call all of it »work«. Do my type design skills exceed e.g. my photography skills? That of course is another question …

What is the relationship between contemporary graphic design and contemporary typefaces? How do you think these two fields affect each other? Do trends in contemporary graphic design emerge from trends in contemporary type design or the other way round? Is it possible to have one without the other?

I would say that a typeface does not necessarily need graphic design, but it seems obvious to me that graphic design needs type.

There was a time when typefaces were designed only for special demands. Today many type designers design (for example) classic text faces or display type, although there are a huge number of pre-existing typefaces with a similar function. What is their motivation?

This is a very funny question, since it seems to imply that somewhere at the beginning of the world we were given a restricted number of typefaces that could be made and once that number was created, there would be no reason for any new ones! As language will keep changing (e.g. linked to technological progress), there will be a need for new forms as well.

Online sources like »myfonts.com«[4] offer an unbelievable number of new typefaces. Most of them appear in big packages with many cuts. Your typefaces, on the other hand, consist mostly of a small number of cuts or even just one.

I am just very slow.

How many cuts does a typeface need? Do you think these are marketing strategies or are they really needed? Do we need »super-families« or is there a limit?

That depends … and probably there is no formula to this. Personally, I don't think the packages you referred to as »super-families« really offer what they claim to offer. Sometimes it seems to me that the only reason for their existence is the fact that it is technologically possible to create them.

1 radimpesko.com

2 OpenType is a current data format for digitalized fonts, developed by Microsoft and Adobe, which overcomes the limits of the usual font formats PostScript (Type 1) and True Type as well as the limited range of characters of PC/MAC (→ S. 164).

3 The Brno International Biennial of Graphic Design took place for the first time in 1963 and now is worldwide one of the oldest events in the area of Graphic Design.

4 MyFonts is a digital font distributor, based in Woburn, Massachusetts, selling fonts through the myfonts.com web site. It was created by Bitstream Inc., launched in September 1999 (during the ATypI conference in Boston), and started selling fonts in March 2000.

Do you have a special target group? Who do you think should use your fonts, and for which projects?

I don't think in terms of »markets« or »target groups« – that would probably be the end for me. I see typefaces as autonomous work. People will find their own way to use them, or they won't. Some things which I did eight years ago got recognised and picked up only just recently. One needs to be patient.

In recent years many young type foundries established themselves in the market, especially from Switzerland and from the Netherlands. What's the reason for that? Do you think this is just a trend, or is it a long term development?

I guess it is partly due to technologies, accessibility and economy. The same thing happens in other fields, too – take music, films or publishing, for example. I don't see why type design should be anything special in this respect.

Secondly, Switzerland and the Netherlands have been traditional leaders of European design since the 1930's, so again it does not surprise me that many young type foundries come from there. On the other hand, borders are progressively getting less important. People from other countries might contribute something new while those traditional leaders might get repetitive or tired of themselves. Who knows …

You are currently still living in Amsterdam. What kind of impact does the city have on your work? Is it important for a young type designer to live in Amsterdam / Lausanne / The Hague …?

The places you live in have a great impact on your life. Amsterdam is an inspiring city – but not especially from a type designer's point of view as there is no big type scene in Amsterdam (maybe that is exactly what made Amsterdam so appealing for me).

Anyhow I have decided to move. Where to is a secret, of course …

Radim Peško was interviewed by David Dusanek and Matthias Dufner.

TIMO GAESSNER

abcdefghijklmnopqrstuvwxyz
ABCDEFGHIJKLMNOPQRSTUVWXYZ
0123456789

Maison Neue Medium

abcdefghijklmnopqrstuvwxyz
ABCDEFGHIJKLMNOPQRSTUVWXYZ
0123456789

Maison Neue Mono

Dare to be naïve.

Maison Neue Bold – 85pt

Thinking is a momentary dismissal of irrelevancies.

Maison Neue Book Italic – 40pt

The Things to do are: the things that need doing, that you see need to be done, and that no one else seems to see need to be done.

Maison Neue Mono – 9pt

Pollution is nothing but resources we're not harvesting. We allow them to disperse because we've been ignorant of their value.

Maison Neue Medium – 12pt

Schrift. Typeface. Maison Neue
Gestalter. Designer. Timo Gaessner
Label. Foundry. milieugrotesque.com
Jahr. Year. 2012

Maison Neue ist die neu gezeichnete Version der früheren Maison-Schriftfamilie. Während die Originalversion auf geometrischen Prinzipien basiert, wurde die Maison Neue nach optischen Aspekten überarbeitet. Neben dem Fokus auf Harmonie, Rhythmus und Lesefluss sind auch aktuelle Display- und Reproduktionstechniken berücksichtigt worden. Das Ergebnis ist eine zeitgemäße, klassische Grotesk mit einem freundlichen Duktus. Die Schriftfamilie umfasst bisher zwölf Schnitte mit erweitertem Zeichensatz und verschiedenen OpenType-Features.

Maison Neue is the thoroughly reworked version of the early Maison typeface family. Whereas the original version was constructed on geometric principles, Maison Neue has now been meticulously redrawn. Paying particular attention to harmony, rhythm and flow, the new typeface also accounts for up-to-date display and reproduction technologies to create a distinctly contemporary grotesque, with a classic touch and a friendly appearance. The whole family contains twelve styles so far; an extended Latin character set and a variety of OpenType features.

Nach seinem Abschluss an der Gerrit Rietveld Academie 2002 gründete *Timo Gaessner*[1] (*1975) in Berlin das Studio »123buero«.[2] Neben seiner Arbeit als Grafik-Designer widmete er sich zunehmend der Schriftgestaltung und entwickelte unter anderem für das türkische Kulturinstitut SALT die Schrift Kralice.[3] 2010 gründete Gaessner zusammen mit Alexander Colby die unabhängige Foundry »Milieu Grotesque«, über die er seine Schriften vertreibt. Der Gestalter hat an der Hochschule für Gestaltung (HfG) in Karlsruhe Typografie unterrichtet und leitet derzeit zahlreiche Workshops.

Was hat Dein Interesse an Schriftgestaltung geweckt?

tg: Ich habe eine Ausbildung zum Schilder- und Lichtreklamehersteller gemacht und in der Berufsschule war ein Teil des Unterrichts Schriftkonstruktion. Mit diesem Hintergrund habe ich in Maastricht an der Kunstakademie mein Studium begonnen, allerdings empfand ich Grafik-Design dort als nicht besonders spannend und wechselte an die Rietveld, um unter anderem bei Gerard Unger[4] zu studieren. Obwohl ich nicht genau wusste, was es bedeutet Schriftgestaltung und Grafik-Design zu studieren, war es ein echter Glücksgriff. Es war sehr interessant, unterschiedliche und teilweise auch sehr krasse Positionen kennenzulernen.

Inwiefern krass?

Sie haben die Haltung vertreten, dass Grafik-Designer jegliche Autorenschaft übernehmen sollten.

Es wurde viel von den Studenten gefordert und ein hoher Leistungsdruck aufgebaut. Man konnte zum Beispiel auch gefeuert werden. Eigentlich hatte ich bis dato nicht wirklich verstanden, dass es den Beruf Schriftgestalter tatsächlich auch gibt. Erst nach dem Studium entwickelte sich der Gedanke, dass man auch davon leben kann. Ich habe mir die Zeit genommen, mich über Jahre in die Thematik einzuarbeiten.

Du hast an der Hochschule für Gestaltung in Karlsruhe unterrichtet. Hat Dich die Erfahrung, die Du als Student gemacht hast, als Lehrender beeinflusst? Was versuchst Du Deinen Studenten mitzugeben?

An der Rietveld gab es eine klare Hierarchie zwischen Lehrer und Student. Diese Hierarchie gibt es bei mir nicht. Im Unterricht versuche ich, aktuelle Themen mit einzuarbeiten und zu moderieren. Aber natürlich beeinflusste mich die Zeit an der Rietveld. Ich fordere meine Studenten auch sehr, lege Wert darauf, dass sie viel arbeiten und nicht den einfachsten Weg suchen. Sie sollen sich mit dem, was sie machen, identifizieren. Das ist vielleicht die stärkste Parallele zur Rietveld. Wenn Du Dich dort nicht mit den Inhalten identifizierst, hältst Du es auch nicht lange aus.

1 timogaessner.de

2 123buero.com
(→ S. 98)

3 Die serifenlose Schrift Kralice wurde 2001 für die türkische Kulturinstitution SALT entwickelt.

4 Gerard Unger
(→ S. 58)

Unterscheidet sich das, was Du von Deinen Studenten erwartest, von Deiner eigenen Arbeitsweise? Wie ist es bei Dir, wenn Du eine Schrift gestaltest? Womit fängst Du an, gibt es eine klare Routine in Deiner Arbeit oder Prinzipien, die immer dazu gehören?

Ein ganz wichtiger Punkt ist, dass ich nicht von Hand arbeite und nicht nach der romantischen Vorstellung mit dem Pinsel da sitze und zeichne. Das habe ich früher gemacht. Heute bin ich mit den digitalen Mitteln so vertraut, dass sie zu meinem bevorzugten Werkzeug geworden sind. Den Studenten lasse ich den Freiraum so zu arbeiten, wie sie es für richtig halten. Mit denen, die noch keine konkreten Ideen haben, mache ich zu Beginn experimentelle Übungen. So ähnlich arbeite ich auch. Es gibt lange experimentelle Phasen. Irgendwann gibt es den Moment, wo man merkt, dass die Idee so weit ist, dass sie umgesetzt werden kann. Diese Phasen können unter Umständen Jahre dauern.

Das Verständnis für Schriften ist ja scheinbar nicht so groß. Kommen trotzdem Leute auf Dich zu und wollen ihre eigene Schrift? Haben sie konkrete Vorstellungen?

Es stimmt, dass die Kunden oft nicht genau wissen, dass eine Schrift auch gestaltet werden kann. Das Bewusstsein entsteht gerade erst. In der Fotografie ist das ähnlich. Sie ist Bestandteil einer Disziplin, die auch konzeptionell begriffen werden kann. Also nicht nur als dekorativer Zusatz, sondern als tragendes Kommunikationsmittel.

Bei der Amentype[5] war es auch ein etwas längerer Prozess. Ursprünglich sollte nur ein bestehendes Corporate Design überarbeitet werden. Mit der Zeit hat sich dann die Idee durchgesetzt, sich primär über die Kommunikationsmittel zu definieren und wir haben drei Schnitte entwickelt.

Das bedeutet, Du benutzt Deine eigenen Schriften gerne?

Immer mehr. Am Anfang habe ich eher ausprobiert. Die NAIV,[6] meine erste Schrift, hatte zum Beispiel einen rein formalen Anspruch. Mittlerweile habe ich kein Interesse mehr an rein formalen Ideen. Es dauert einfach eine gute Zeit, bis sich das Auge und der Geschmack entwickelt haben. Es ist immer wieder spannend, meine Schriften gedruckt eingesetzt zu sehen. In Tokio habe ich einmal meine Schrift in einem riesigen Supermarkt voller Magazine entdeckt. Natürlich gibt es manchmal auch Anwendungen, die mir nicht gefallen. Man hat das Gefühl, manche Leute verstehen das Konzept nicht. Generell sehe ich meine Schriften aber super gerne angewandt, auch wenn das nicht immer meinem Geschmack entspricht.

Wie lange arbeitest Du durchschnittlich an einer Schrift? Bist Du sehr perfektionistisch veranlagt?

Qualität ist mir sehr wichtig. Ich möchte immer mein ganzes Wissen einbetten und das Wissen wächst stetig. Je mehr ich mich damit beschäftige, desto ausgereifter werden die Schriften. Der Chapeau[7] merkt man an, dass sie vier Jahre alt ist, dagegen ist die Maison Neue fast »perfekt«.

Die Maison Neue ist perfekt? Allerdings gab es ja schon zwei Jahre vor der Veröffentlichung die Maison. Was hat Dich dazu veranlasst, die Schrift erneut zu überarbeiten?

Beide Schriften basieren auf unterschiedlichen Konzepten. Die Maison orientiert sich an alten Groteskschriften. Die Idee war, einen groben Duktus zu entwickeln, der an frühe Akzidenzschnitte erinnert, basierend auf einem Raster mit entsprechend wenig optischen Korrekturen. Wie eine Schrift aus dem frühen

5 Die Amentype wurde 2011 für die Start-up-Internetplattform Amen entwickelt. Die weboptimierte Groteskschrift umfasst drei Schnitte.

6 Die Schrift NAIV ist eine Grotesk mit abgerundeten Endungen und angedeuteten Skriptelementen.

7 Die Schrift Chapeau wurde von Johnny Cashs Briefen an seine Fans inspiriert. Cashs Schreibmaschine dient als Basis der geometrischen und proportional angepassten Grotesk. Die Chapeau wird durch die eigene Foundry Milieu Grotesque vertrieben.

zwanzigsten Jahrhundert. Ich habe versucht, die Idee subtil und fein umzusetzen. Dennoch wurde oft kritisiert, dass die Schrift zu grob sei. Sie wurde als nicht »perfekt« wahrgenommen. Dabei war es das Prinzip der Schrift. Daraufhin habe ich die Maison Neue gestaltet. Die alte Maison war eine tolle Basis. Dadurch war es dann relativ einfach, sie unter optischen Aspekten zu überarbeiten. Hierfür habe ich mich zuerst auf die Harmonie der Buchstaben und dann auf das Zusammenspiel untereinander konzentriert. Ich wollte eine moderne Schrift, die eine breite Anwendungsmöglichkeit bietet.

Hättest Du nicht auch mal Lust eine Serifenschrift zu gestalten?

Doch auf jeden Fall. Ich arbeite schon länger an einer Serifenschrift. Die ist einfach nur noch nicht so weit, dass ich sie veröffentlichen kann. Das liegt aber auch daran, dass ich mich schon viel länger mit Groteskschriften beschäftige. Serifen sind ein neues Thema, in das ich mich noch einarbeiten muss. Das Auge muss sich erst einmal daran gewöhnen. An der Schrift arbeite ich seit etwa zwei Jahren, ich glaube aber, es dauert noch mindestens drei Jahre, bis sie fertig ist.

Was sagst Du eigentlich zu diesen ganzen Type Foundries, die momentan aus dem Boden sprießen? Glaubst Du, dass es sich hierbei um einen Trend handelt oder ist das eine nachhaltige Entwicklung?

Was mich ein bisschen nervt, sind Gestalter, die schnell mal ein paar Buchstaben zusammenklicken, ein Plakat daraus machen und dann versuchen beides zu verkaufen. Ich glaube, dass dieses Konzept wenig nachhaltig ist. Andererseits gibt es viele junge, sehr gute Leute in diesem Bereich. Ich finde es schade, dass es so wenig gute deutsche Nachwuchs-Foundries gibt. Schriftgestaltung wird an den Hochschulen einfach zu wenig unterrichtet. An der HfG Karlsruhe war ich seit zehn Jahren einer von Wenigen.

Was war Deine Motivation, eine Foundry zu gründen?

Ich hätte es auch gut gefunden, wenn meine Schriften bei einer anderen Foundry untergekommen wären, jedoch gab es keine passende. Meine erste Schrift, die NAIV, ist bei Gestalten und Fontshop erschienen. Dort bekommt der Gestalter nur einen kleinen Anteil. Außerdem gibt es weder Qualitätsmanagement noch eine angemessene Beratung. Uns schien es, als gäbe es nur wenige Foundries, die eine Haltung vertreten, für die sie sich auch einsetzen. Also haben wir einfach eine gegründet. Das macht trotz der vielen Arbeit eine Menge Spaß.

Hattest Du während deines Studiums Vorbilder?

Rietveld beeinflusst einen auf jeden Fall. Während ich dort war, habe ich die Schule nicht gemocht, aber hinterher hat alles Sinn ergeben. Linda van Deursen[8] vom Büro Mevis & Van Deursen schätze ich auch heute noch. Was deutsche Gestalter angeht, finde ich natürlich Otl Aicher[9] toll, obwohl er ein Dogmatiker ist. Weniger seine Arbeiten, mehr seine persönliche Art, seine Konzeption und seinen philosophischen Ansatz.

Welche Entwicklung wird die Typografie in den nächsten Jahren Deiner Meinung nach nehmen? Kannst Du die Entwicklung Deiner Arbeit abschätzen?

Ich glaube, dass Design keine Dienstleistung im klassischen Sinne mehr sein wird. Die Technik wird bald so ausgereift sein, dass es kein besonderes Know-how mehr braucht, um sauber umsetzen zu können. Hoffentlich werden Desig-

8 Linda van Deursen (*1961) ist eine niederländische Grafikerin. Während ihres Studiums an der Gerrit Rietveld Academie lernte sie Armand Mevis kennen. Gemeinsam gründeten sie 1987 das Büro Mevis & Van Deursen. Neben der Arbeit an Erscheinungsbildern und Publikationen für kulturelle Institutionen unterrichtet Linda van Deursen an der School of Art der Yale University.

9 Otl Aicher, eigentlich Otto Aicher (1922–1991), deutscher Designer, Pionier des Corporate Designs und maßgeblich an der Entwicklung des Erscheinungsbildes u.a. für die Lufthansa, das ZDF, die Dresdner Bank oder Braun beteiligt. 1988 veröffentlichte er seine Hybridschrift Rotis → S. 74.

ner dann mehr die Rolle des Moderatoren oder Kommunikationsberaters ausfüllen. Ich glaube auch, dass viele Menschen keine Lust mehr auf große Firmen und industrialisierte Produkte haben und sich daher lieber selbst behelfen. Ich persönlich möchte mich auf wenige gute Projekte als Grafik-Designer konzentrieren, mehr Zeit für Schriften haben und gerne auch mehr unterrichten. Die Zeit an der HfG Karlsruhe, in einem unkommerziellen Umfeld, hat Spaß gemacht.

Das Interview mit Timo Gaessner
führten Anna Alexander,
Lisa Grünwald und Bahar Hasan.

Tensegrity

90 m²

Biosphère

Otisco

Dymaxion

Maison Neue Demi Italic
Maison Neue Book Italic
Maison Neue Bold

Maison Neue Mono Italic
Maison Neue Medium

After his degree at the Gerrit Rietveld Academie (in 2002), *Timo Gaessner*[1] (*1975) started a graphic design studio in Berlin *»123buero«*.[2] In addition to his work as graphic designer, he was increasingly engaged in type design, and, among other things, he developed a font called Kralice[3] for SALT, a Turkish cultural institute. In 2010, he established an independent foundry »Milieu Grotesque«, where he has been marketing his own fonts. Since 2011, Timo Gaessner has been teaching the subject of typography at the Hochschule für Gestaltung (HfG) in Karlsruhe and has been running numerous workshops.

How did you become interested in type design?

tg: I learned to produce signboards and neon signs, and some of the lessons at the vocational college included type design. With this background knowledge, I began my studies at the Academy of Arts in Maastricht. However, I didn't find graphic design very interesting there, and for this reason I went to Gerard Unger[4] at the Rietveld Academie. Though in fact I didn't really know what was involved in studying type design and graphic design, this was a real stroke of luck. It was quite fascinating to work with people who were truly experienced in this field and sometimes had rather extreme ideas about graphic design.

What do you mean by »extreme«?

They were of the opinion that graphic designers should have the intellectual authorship over all activities in a cultural society. At that time, the students had to meet very high demands, and there was a lot of stress. So, for example, students could be sacked. And I myself, for instance, hadn't really understood up to that time that there actually was such a profession as type designer. It wasn't until after I had graduated that I realised one could actually make a living from that, and I took my time getting acquainted with the subject.

You worked as an instructor at Karlsruhe University of Arts and Design. Did your experience as a student influence your teaching? What do you want your students to take away with them when they leave?

At Rietveld, there was a clear hierarchy between teachers and students. I don't agree with that. In teaching, I try to discuss open questions and find answers. But of course my time at the Rietveld Academy has had its influence upon me.

I also demand a lot from my students. I expect them to work hard and don't settle for the easiest way. They should identify themselves with their work – this is probably the greatest similarity to Rietveld. There, if you don't identify yourself with your work, you won't last very long.

Is there any difference between the work you expect from your students and your own approach to your work? How do you go about creating a typeface? What is your first step, do you have a definite routine or any basic principles?

It is very important to note that I don't work by hand; I don't have this romantic notion that I need to use a brush for my drawings. The new digital media allow me to work very fast and so they have become my preferred technique. I give my students leeway to work in their own way. Initially, I give some experimental exercises to those who don't have any definite ideas yet. This is similar to my own way of working. There's a very long experimental phase of trial and error until you realize that the time has come to put your ideas into practice. Sometimes this process can take years.

It seems that type isn't sufficiently appreciated. In spite of that, are you being contacted by people who want to have their own fonts? Have they got any concrete ideas?

It's true that people don't really know that you can in fact design a typeface. It is only now that you notice an awareness about that. It's similar to photography, it is part of a discipline which can also be considered as conceptual, i.e. not as a decorative »extra«, but as a means of communication in its own right.

Amentype[5] also was rather a lengthy process. Initially it was only a question of reworking an existing corporate design. Over time, however, the idea was born to define oneself through the primary communication media and we developed three type styles.

Does this mean that you like to use your own fonts?

Yes, more and more. At the beginning, I tried out a lot of things. NAIV,[6] for instance, which was my first font, was designed for merely formal reasons. But now I have lost interest in those purely formal ideas. It takes time to develop your perception and taste. It's exciting to see my fonts in publications. When I was in Tokyo, I discovered one of my fonts in a huge supermarket full of magazines. But there are also those applications which I don't like. You get the feeling that some designers don't understand the idea behind a certain font. In general, however, I very much like to see my fonts in use, even if it isn't always exactly as I imagined it.

1 timogaessner.de

2 123buero.com
(→ S. 103)

3 The sans-serif typeface Kraliçe was developed 2001 for the Turkish cultural institution SALT.

4 Gerard Unger
(→ S. 62)

5 The Amentype was developed in 2011 for the startup internet platform Amen. The web optimized Grotesque consists of three styles.

6 The typeface NAIV, developed in 2005, is a Grotesque with rounded ends and hints of script elements.

How long on average does it take you to create a font? Are you a perfectionist?

Quality is very important to me. I always want to apply my total knowledge – and this is growing continually. The more time I spend over a typeface, the more perfect the fonts will be. It's clearly visible that the Chapeau[7] is already four years old. Maison Neue, on the other hand, can be considered almost perfect.

Maison Neue is perfect? But the Maison Typeface was already there, two years before the Maison Neue was published. What made you rework this font?

Both fonts are based on different ideas. Maison derives its orientation from old sans-serif typefaces. I basically wanted to develop a characteristic style resembling early cuts in the field of job printing, and I wanted to achieve this by making just a few optic corrections – like a font of the early twentieth century. I tried to realize this in a rather subtle and smooth way. Even so, the font was often criticized for being too rough. It was considered not to be »perfect«, although that was exactly the principle behind this font.

As a result of it, I created Maison Neue. The old Maison was a good base for its development, and with this it was relatively simple to rework this font, considering current optical aspects. First of all, I concentrated on creating harmonious letters, and then I focused on their interaction. I wanted to have a modern typeface with a wide range of applications.

Wouldn't you like to create an Antiqua typeface as well?

Yes, for sure! I've been working at a serif typeface for quite a while now. It's just not yet ready for publication, which is also due to the fact that I have been dealing with sans-serif typefaces much longer. The subject of serif typefaces is something new for me; I am still getting acquainted with it. My eyes still have to get used to these kinds of typefaces. I have been working on this font for about two years now, but I think it will take at least three more years to finish it.

What do you think about all these type foundries which have mushroomed over the past few years? Do you think this is just a trend or is it an ongoing development?

I find it annoying that there are people who fancy themselves as typographers or type designers just because they can put ten letters together or create a poster and then even try and sell both of them. I don't think that this is a concept which will last. On the other hand, you find a great many young and quite excellent people in this field. It is rather regrettable that there are so few new foundries of good quality in Germany. The topic of type design is not sufficiently covered at universities; at Karlsruhe University of Arts and Design, over a ten-year period, I was one of very few people teaching this subject.

What gave you the idea to set up a foundry?

I wouldn't have minded if my fonts had gone to another foundry, but we didn't find a suitable one. My first fonts, NAIV and Maison, were published by Die Gestalten and Fontshop. Here, the designer only receives a small proportion, besides, there is neither quality management nor any consulting.

To us, many foundries don't have any conceptual background, nothing they stand for. That is the simple reason why we established our own foundry. And although it is a lot of work, it is also a lot of fun.

Did you have any role models during your studies?

Rietveld certainly had an influence. I didn't particularly like it while I studied there, but afterwards everything made more sense. I'm also a great admirer of Linda van Deursen[8] of the Mevis & Van Deursen studio. Among the German designers, I certainly appreciate Otl Aicher;[9] even though he is a dogmatist. Actually I'm not really enthusiastic about his work, but rather more about his personal disposition, his way of designing, and his philosophical approach.

What do you think about the development of typography over the next few years? How do you see the future development of your own work?

I expect that in future design will no longer be a service in the traditional sense. Before long the technology will be sufficiently developed so that you won't need a particular know-how any more in order to implement your ideas. I hope that the designer will then assume the role of a presenter and a communication consultant. The trend goes towards individualism; people aren't any longer keen on big enterprises and industrialized products, and that's why the idea of »Do it yourself«. will catch on more and more. Personally, I would like to concentrate on just a few good projects as a graphic designer, I would also like to have more time for type design and for teaching. I enjoyed my time at the Karlsruhe University of Arts and Design.

Timo Gaessner was interviewed by Anna Alexander, Lisa Grünwald, and Bahar Hasan.

7 The typeface Chapeau was inspired by Johnny Cash's letters to his fans. The typeface Cash's Schreibmaschine (Cash's typewriter), developed in 2009, is the basis of the geometrical and proportionally adapted Grotesque.

8 Linda van Deursen (*1961) is a Dutch graphic designer. While studying at the Gerrit Rietveld Akademie she met Armand Mevis, and together they launched the studio Mevis & Van Deursen in 1987. Apart from working on identities and publications for cultural institutions, Linda van Deursen is teaching at the School of Art at Yale University.

9 Otl Aicher, (Otto Aicher, 1922–1991), German designer, pioneer of corporate design and essentially responsible for the corporate identity of Lufthansa, ZDF, Dresdner Bank and Braun, among others. In 1988 he published his hybrid typeface Rotis → S. 77.

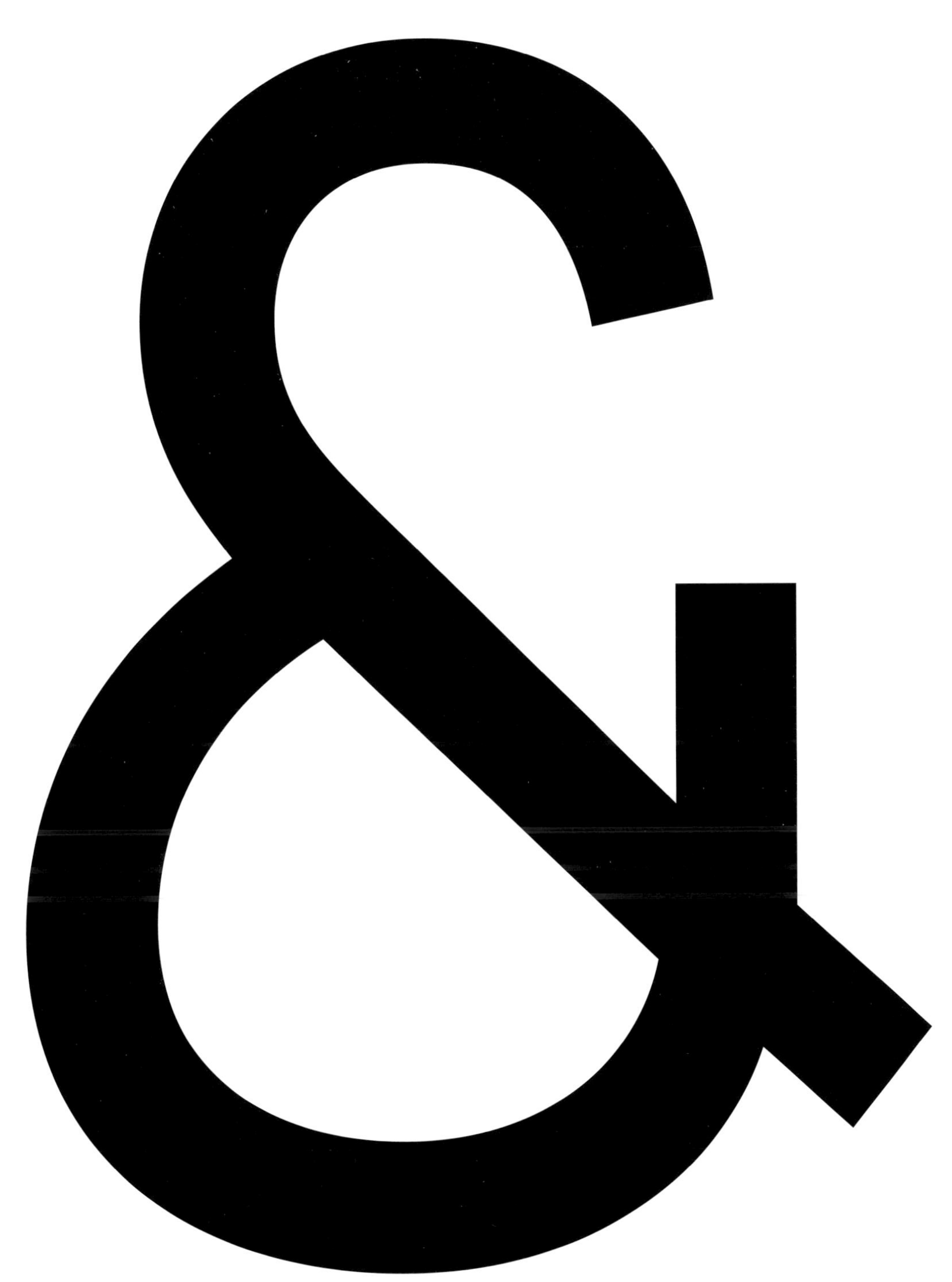
MAISON NEUE

IMPULSIVE, VIBRANT, MOODY & DELICATE

ABCDEFGHIJ
KLMNOPQRS
TUVWXYZ

MeM Impulsive

ABCDEFGHIJ
KLMNOPQRS
TUVWXYZ

MeM Vibrant

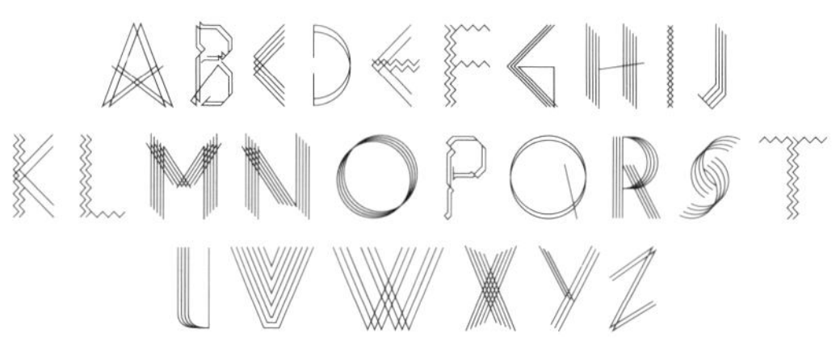

MeM Moody

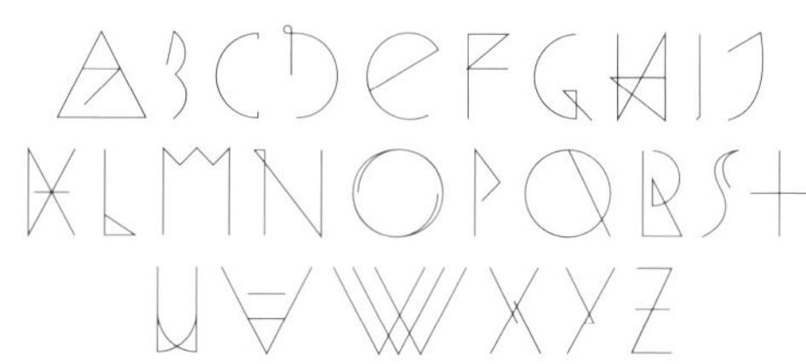

MeM Delicate

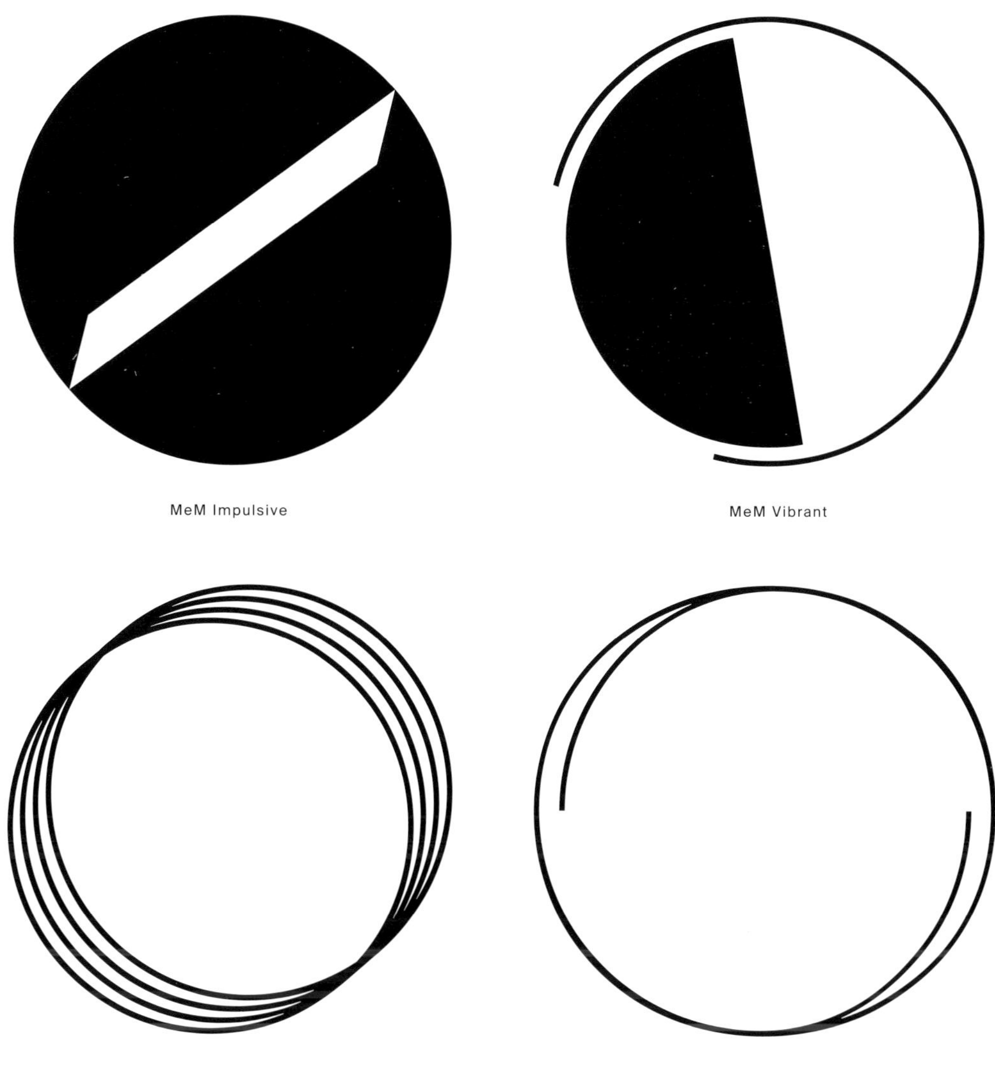

MeM Impulsive

MeM Vibrant

MeM Moody

MeM Delicate

Schrift. Typeface. MeM
Gestalter. Designer. Jakob Runge & Elena Schädel
Label. Foundry. 26plus-zeichen.de
Jahr. Year. 2012

Das experimentelle Versal-Alphabet MeM rückt die Lesbarkeit in den Hintergrund und lässt die Grenzen zwischen Grafik-Design und Schriftgestaltung verschwimmen. Die vielen Alternativzeichen der Schrift und die extremen Stile der einzelnen Glyphen erzeugen eine lebendige Spannung aus geometrischer Anmutung und verspielten Formen, wie eine kontrastreiche Vermischung von ausdrucksstarken und feinteiligen Elementen – jeder Buchstabe ist ein Stück Grafik für sich.

Was wie eine Aneinanderreihung von Vektorformen wirkt, ist ein durchdachtes Schriftsystem: ein automatisches Shuffle-Prinzip mixt in den gängigen Designprogrammen die ohnehin verteilten Schriftgewichte und verhindert die Wiederholung gleicher Zeichen. Im gleichen Maße wie die Schrift durch OpenType-Features aufgemischt wird, ist sie auch gezielt anwendbar: Durch Stylistic Alternatives können kräftige und leichte Zeichen voneinander getrennt oder auch komplett in einen einzelnen der vier Schriftschnitte zerlegt werden.

In the experimental upper-case alphabet MeM readability is of secondary consideration, here the borderline between graphic design and typography is blurred. The many alternative signs of a type and the extreme styles of the glyphs create a lively coexistence of geometric impression and playful form, like a contrasting mixture of expressive and delicate elements – each letter its own graphic creation.

What at first looks like a series of vector forms, is in fact a well thought-out system: an automatic shuffle principle, as used in the usual design programmes, mixes the already distributed type weights and prevents the repetition of identical signs. And in the same way as the type is mixed up by OpenType features, it can also be used: using Stylistic Alternatives, bold and light characters can be separated from each other, they can even be separated completely into one of the four type styles.

Der Kommunikationsdesigner und Mitbegründer der Präsentationsplattform studentischer Schriftentwürfe »26+«[1] *Jakob Runge*[2] (*1985) ist als selbstständiger Schrift- und Corporate-Designer in München tätig. Nach seinem Studium an der FH Würzburg absolvierte er mit der Franziska[3] seinen von Albert-Jan Pool[4] betreuten Master an der Muthesius Kunsthochschule Kiel. Zu seinen erfolgreichen Veröffentlichungen zählen die Sinews Sans (Gestalten Fonts), die TJ Evolette A (YouWorkForThem; in Zusammenarbeit mit Timo Titzmann) und die *MeM* (YouWorkForThem & HypeForType; in Zusammenarbeit mit *Elena Schädel*[5]).

Du bist also professioneller Schriftgestalter …

jr: Ja, ich sehe mich als Schriftgestalter. Ob ich mich tatsächlich schon so bezeichnen darf, ist natürlich die andere Frage. Von Schriftgestaltung zu leben, ist nicht so einfach: Ich schätze, in Deutschland gibt es vielleicht fünfzig bis hundert Leute, die allein von Schriftgestaltung leben können. Für den Rest ist sie nur Zubrot oder Hobby.

Was zeichnet Deine Schriften aus?

Was meine bisherigen Schriften angeht, so kann man meinen Schaffensbereich im Prinzip in zwei Bereiche aufteilen: Auf der einen Seite sind das sehr funktionale, kundenorientierte Schriften, wie zuletzt die Franziska, auf der anderen Seite sehr experimentelle Schriften, wie z.B. die MeM, bei denen man in der Entwicklungsphase noch gar nicht abschätzen kann, ob sie später nutzbar sein werden.

Wie lange arbeitest Du an experimentellen Schriften wie der Evolette[6] oder der MeM?

Diese Schriften waren tatsächlich zwei absolute »Schnellschüsse«. Im Fall der Evolette ließen wir uns einfach nur von aktuellen Grafik-Trends leiten. Zunächst stand der Spaß im Vordergrund. Erst später haben wir beschlossen, die Schrift weiter auszubauen und zu verkaufen. Der Vorteil einer Display-Schrift ist natürlich immer, dass die Entwicklungszeit relativ kurz und ihr Stil bei Fertigstellung noch aktuell ist.

Wie kamst Du dazu, mit der Franziska eine moderne Textschrift zu Deiner Master-Arbeit zu machen?

Ich sehe das wie eine Art Gesellenstück – eine Arbeit, die mir sehr viel abverlangt und mir neue Wege eröffnet. Zunächst war es nur eine Semesterarbeit, erst mit der Zeit wurde mir aufgrund der wachsenden Anzahl der Schnitte die Dimension des Projekts bewusst, so dass ich

1 26plus-zeichen.de

2 jakob-runge.de

3 Halb Antiqua, halb Egyptienne vereint die Franziska eine gute Lesbarkeit in Fließtexten mit interessanten Charakteristika in großen Punktgraden.

4 Albert-Jan Pool (*1960) ist niederländischer Type Designer. Die FF DIN und FF OCR-F waren seine ersten Schriftprojekte. Darüber hinaus schuf er die Jet Set Sans, C&A InfoType, DTL HEIN GAS und HEM Headline Corporate Typefaces. Er unterrichtete Type Design an der Muthesius Kunsthochschule.

5 Elena Schädel (*1985) erhielt ihr Diplom für Kommunikations- und Informations-Design an der FH Würzburg. Sie gründete das Grafik-Büro Elena Schädel Design (elenaschaedel.de) und ist als selbstständige Grafik-Designerin in Melbourne (Australien) tätig. Ihre Master-Thesis untersucht, wie sich das Menschenbild in den Humanwissenschaften seit der digitalen Revolution veränderte. Schädel veröffentlichte die Schrift MeM (YouWorkForThem & HypeForType; in Zusammenarbeit mit Jakob Runge).

6 Die geometrisch konstruierte Versalschrift TJ Evolette A zeichnet sich durch extravagante Varianten an Alternativzeichen für das gesamte Alphabet sowie überraschend gestaltete Sonderzeichen aus.

mich entschloss, auch meine Master-Thesis auf der Franziska aufzubauen. Im Endeffekt habe ich sehr viel durch den Prozess der Entwicklung der Franziska über Schriftgestaltung im Allgemeinen gelernt.

Wo siehst Du Deine Zukunft im Bereich des Type Design?

Um ganz ehrlich zu sein, muss ich sagen, dass ich mir eigentlich erst nach meiner Master-Thesis eingestehen konnte, dass ich mich vollkommen auf Schriftgestaltung spezialisieren möchte. Ich war zuvor immer davon ausgegangen, dass ich mich nebenbei noch mit Bereichen wie Webentwicklung oder Editorial Design beschäftigen muss – auch um Geld zu verdienen. Insofern kann ich im Moment noch gar nicht so genau sagen, wo ich mich in Zukunft im Bereich der Schriftgestaltung sehe, nur dass ich meine Brötchen mit Buchstaben verdienen möchte.

Wie sehen Deine weiteren Pläne aus?

In der zweiten Jahreshälfte 2013 habe ich vor, mich als Schriftgestalter selbstständig zu machen. Ich werde also meine Schriften bei verschiedenen Type Foundries unterbringen und kleinere Corporate Fonts für Kunden entwickeln. Abgesehen davon widme ich mich dem Bereich Logo-Design, in dem Schriftgestaltung meistens eine große Rolle spielt.

Du hast Grafik-Design studiert und arbeitest jetzt als Schriftgestalter. Ist es Deiner Meinung nach wichtig, die andere Seite zu kennen? Und inwiefern beeinflussen Trends im Grafik-Design Deine Entwürfe?

Zunächst einmal würde ich sagen, dass ein Schriftgestalter zum Teil immer auch Grafiker sein muss. Man produziert ein Werkzeug, das später nutzbar sein muss. Wenn ich beispielsweise eine Schrift gestalten möchte, die im Web gut funktioniert, muss ich wissen, welche technischen und ästhetischen Anforderungen an aktuelle Webfonts gestellt werden. Zugleich ist es wichtig, Grafik-Trends zu kennen, um den Markt mit gefragten Schriften versorgen zu können.

Wie wichtig ist dieser technische Aspekt für Dich?

Man kann bei der Entwicklung unglaublich viele Features in eine Schrift packen. Tatsächlich merken jedoch viele Gestalter gar nichts davon, weil sie die Möglichkeiten von OpenType[7] nicht kennen oder einzusetzen wissen. Deshalb kann man als Schriftgestalter natürlich auch sagen: Okay, ich spare mir das und mache einen Alternativschnitt mehr. Damit können mehr Leute etwas anfangen.

Dennoch finde ich OpenType-Programmierung sehr spannend, weil sie der Schriftgestaltung einen neuen Aspekt hinzufügt und dem Gestalter die Möglichkeit eröffnet, seine Schriften auf einer weiteren Ebene mitzugestalten.

Wo siehst Du die Aufgabenstellung des jungen Type Design? Geht es um Neuerungen oder nur darum, die Tradition fortzuführen?

Das ist eine gute Frage. Im Type Design besteht das Dilemma, dass man sich nicht allzu weit von den gewohnten Formen entfernen kann. Es gibt also wenig Raum für große Innovationen. Ich glaube aber, dass talentierte Schriftgestalter aufgrund der technischen Produktionsmöglichkeiten heutzutage wesentlich hochwertigere und umfangreichere Schriften produzieren können als früher. Man kann z.B. in derselben Zeit mehr Korrekturschleifen durchlaufen.

Du hast »26+« mitbegründet. Wie wichtig ist Nachwuchsförderung?

Im Bereich Schriftgestaltung ist das schwer, weil er ohnehin sehr speziell ist und sich relativ wenige Leute dafür interessieren. An erster

7 OpenType ist ein aktuelles Datenformat für digitalisierte Satzschriften, von Adobe und Microsoft entwickelt, welches die Grenzen der weit verbreiteten Font-Formate PostScript (Type 1) und TrueType sowie den limitierten Zeichenvorrat von PC/MAC überwindet (→ S. 142).

Stelle stehen daher Austausch und Vernetzung. Das schafft neue Motivation, da man mehrere Jahre am Ball bleiben muss, um qualitativ hochwertige Schriften erstellen zu können. Insofern leistet 26+ als Präsentationsplattform für Schrift-Ideen und als Forum für junge Gestalter einen Beitrag dazu, auch wenn durch die Fülle an Information im Internet inzwischen der Einstieg in das Schriftgestalten deutlich leichter geworden ist.

Wichtig finde ich bei der Ausbildung auch die Vermittlung handwerklicher und historischer Komponenten. Dadurch, dass man mit Programmen wie FontLab »mal eben schnell etwas kopieren kann« wird oft vergessen, dass es bei Schriftgestaltung um mehr geht als nur um die technische Umsetzung von Formen.

Warum sind die »Klassiker« noch immer etabliert?

Das liegt zum einen an der Verfügbarkeit: Wenn man erst einmal längere Zeit als Designer tätig ist, kennt man bestimmte Klassiker und nutzt sie immer wieder gerne – oft auch, weil man sie bereits auf dem Rechner hat. Ansonsten bleibt zur Erprobung im Fließtext meist nur die Möglichkeit, mit nicht lizenzierten Schriften zu arbeiten. Im Zweifel nimmt man daher meist die Schriften, die man schon hat.

Zum anderen glaube ich, dass sich in der Schriftgeschichte zumindest im lateinischen Bereich seit dem 15. Jahrhundert tatsächlich nicht mehr allzu viel getan hat. Die Grundformen der Glyphen sind immer noch gleich. Insofern denke ich, dass die Klassiker Schriften sind, die einfach gut gemacht sind und die sich langfristig durchgesetzt haben.

Du sagst selbst, dass sich an den Schriften heutzutage nicht mehr viel ändert. Trotzdem erscheinen jedes Jahr tausende von neuen Schriften. Wie erklärt sich das?

Hier spielt das Internet eine entscheidende Rolle: Wie man Schriften digital erstellen kann, lässt sich leicht herausfinden. Neue, einfach zugängliche Software erleichtert die Arbeit dabei enorm. Und aufgrund des Erfolgs von Social Media lassen sich Schriften auch relativ leicht vermarkten und verkaufen. Daher ist die Anzahl der Neuerscheinungen gestiegen. Streng genommen könnte man heute aufhören, neue Gestaltungsansätze zu suchen. Das betrifft Schriftentwürfe ebenso wie z.B. Layouts von Büchern. Alles wurde schon einmal gemacht. Dennoch möchte man öfter mal was Neues sehen – etwa eine Schrift, die ein wenig anders ist, mit der man sich individuell abgrenzen kann.

Gibt es eine Schrift, die Du überhaupt nicht magst?

Die Calibri.[8] Nicht etwa, weil sie eine schlechte Schrift wäre, im Gegenteil! Sie ist zurzeit einfach die Schrift, die beweist, dass ihr Anwender typografisch keine Überlegung getroffen und stattdessen die Standardschrift aus Windows belassen hat.

Das Interview mit Jakob Runge führten Matthias Dufner und David Dusanek.

8 Calibri ist Teil einer Reihe neuer Schriftarten, die mit Microsoft Windows Vista eingeführt wurden, und ist darüber hinaus auch in Microsoft Office seit der Windows-Version 2007 beziehungsweise der Mac-OS-X-Version 2008 enthalten. Hier löst sie Verdana als Standard-Sans-Serif-Schrift ab.

»ES GIBT
WENIG
RAUM FÜR
GROSSE
INNOVA-
TIONEN.«

Jakob Runge
→ S. 161

The communication designer and co-founder of the platform for Student Type Design »26+«[1] *Jakob Runge*[2] (*1985) is working as a freelance designer for type and corporate design in Munich. After graduating at the University of Applied Sciences, Würzburg, he created Franziska[3] in the course of his Master Degree under Albert-Jan Pool[4] at the Muthesius Kunsthochschule Kiel. Among his successful publications are the Sinews Sans (Gestalten Fonts), the TJ Evolette A (YouWorkForThem; in cooperation with Timo Titzmann), and the *MeM* (YouWorkForThem & HypeForType; in cooperation with *Elena Schädel*[5]).

You are a professional type designer …

jr: Yes, I consider myself a type designer. However, whether I can really be called one, that's of course another question. It's very difficult to design typefaces for a living. I guess there are perhaps fifty to one hundred persons in Germany who can exclusively live by type design. For the others, type design is just an extra income or a hobby.

What characterizes your fonts?

As far as my previous fonts are concerned, my field of activity can actually be divided into two sectors. On the one hand, there are very functional, customer-oriented fonts, for example, the recently developed Franziska. On the other hand, there are very experimental fonts, such as MeM. When the latter are developed you can't even know whether they will be useful later.

How long does it take you to create experimental fonts like Evolette[6] or MeM?

Actually these two fonts were really »shot from the hip«. As far as Evolette is concerned, we quite simply found inspiration in current graphic trends. At first it was all about having fun. It was only later that we decided to continue with its development and to sell it. Of course it's always favourable that it takes a relatively short period of time for the development of a display typeface and that this typeface is still up-to-date when it's ready.

How did you get the idea to use Franziska, a modern body text font, as the subject of your Master thesis?

To me, it was some kind of final test; I mean it was a very demanding task which enabled me to break new ground. Initially it was only a term paper. However, after a while the increasing number of cuts made me aware of the dimension of the project. So I decided to use Franziska as a basis for my master's thesis, too. Summing it all up, I would say that I learnt a great deal.

How do you see your future in the field of type design?

To tell you the truth, actually it was only after my Master thesis that I could admit to myself that I wanted to specialize completely in type design. Actually I have always been of the opinion that I have to engage in other fields like Web development or editorial design, too – and this also in financial terms. For this reason, it's momentarily difficult to say precisely where I can see myself in the field of type design in the future.

What are your further plans?

I'm planning to start my own business in the field of type design for the second half of 2013. So I will try to supply several type foundries with my fonts and will develop basic corporate fonts for my customers. Moreover, I will apply myself to the field of logo design in which type design often plays an important role.

You studied graphic design and now you are working as a type designer. Do you think that it is important to know the other field, too? How do trends in graphic design influence your work?

First of all, I think that a type designer should also be a graphic designer to some extent. You produce a tool which has to be usable afterwards. For example, if I want to produce a font which functions well on the Web, I must know about the technical and aesthetic requirements which are imposed on current web fonts. At the same time, it's important to know graphic trends: This can enable you to supply the market with fonts which are sought after.

How important is this technical aspect to you?

It always depends. When you develop a font, you can equip it with an unbelievable number of features. In fact, many designers aren't aware of that because they just don't know about the possibilities of OpenType,[7] or they don't know how to use them. Instead, a font designer could of course also say: Okay, I can save myself this effort and make one more alternative character. So more people will be able to handle it.

In spite of that, I find OpenType programming very fascinating because it adds a new aspect to type design and enables the designer to participate in the design of his fonts on another level.

1 26plus-zeichen.de

2 jakob-runge.de

3 Partly serif and slab serif typeface, Franziska unites good legibility in continuous texts with interesting characteristics in large point sizes.

4 Albert-Jan Pool (*1960) is a Dutch professional type designer. FF DIN and FF OCR-F were among his first typeface design projects. Pool also created the Jet Set Sans, C&A InfoType, DTL HEIN GAS and HEM Headline corporate typefaces. Pool has been teaching type design at the Muthesius Kunsthochschule.

5 Elena Schädel (*1985) got her Diploma in Communication and Information Design at the University of Applied Sciences in Würzburg. She established the Graphic Studio Elena Schädel Design (elenaschaedel.de) in Melbourne, Australia, where she is working as a freelance Graphic Designer. Her Master thesis dealt with the question of how the digital revolution has changed the Image of Humanity in the Human Sciences. In cooperation with Jakob Runge, Schädel published the typeface MeM (YouWorkForThem & HypeForType).

6 The geometrically constructed capital typeface »TJ Evolette A« stands out through extravagant variants in its alternative characters for the entire alphabet, as well as some surprising special characters.

7 OpenType is a current data format for digitalized fonts, developed by Microsoft and Adobe, which overcomes the limits of the previous font formats PostScript (Type 1) and True Type as well as the limited range of characters of PC/MAC (→ S. 146).

Where do you see the tasks of young type design? Does young type design focus on innovations, or does it only want to carry on a tradition?

That's a good question. The problem in the field of type design is that you can't depart too far from the common letter forms. There's little room left for big innovations. However, I think that, due to technical possibilities in the field of production, today's type designers, if they are talented, can produce fonts which are much better and more comprehensive than they have ever been before because, among other things, there can be more correction loops within a certain time period.

You co-founded 26+. How important is the promotion of young designers to you?

Promotion is difficult in the field of type design because it's a very specific field anyway, and only relatively few people are interested in it. That's why exchange and networking have top priority. They can create new motivation considering the fact that the creation of high quality fonts requires several years of work. As a platform for the presentation of typographic ideas and a forum for young designers, 26+ certainly supports promotion, even if the access to this field has now become considerably easier by social media.

To me, it's important that professional training also deals with craftsmanship and historic elements. Programmes like FontLab, which enable you »to copy something en passant«, make us often forget that type design should be more than just a technical realization of letter forms.

Why are »traditional« fonts still commonly used?

On the one hand, it's because they are available. If you have been working as a designer for some time, you just know some of these fonts, and you like to use them. This is often due to the fact that you already have them on your computer. Otherwise, if you want to try something out in running text, you are often restricted to unlicensed fonts. For this reason, in case of doubt, you often use the fonts you already have.

On the other hand, I actually believe that there haven't been any significant changes in the history of fonts since the 15th century – at least in terms of Latin fonts. The basic forms of glyphs are still the same. In this respect, I think that traditional fonts are just typefaces of good quality which gained acceptance in the long run.

You are of the opinion that fonts don't change a lot these days. However, thousands of new fonts flood the market every year. How did that come about?

The Internet plays a decisive role. It's easy to find out about the digital production of fonts. New software, which is easily accessible, facilitates this work, and thanks to flourishing social media it's relatively easy to bring typefaces to market and sell them. That's why the number of new fonts has increased. Strictly speaking we could now quit looking for new design ideas. This doesn't only concern the design of fonts but also the layout of books. Everything has been done before. However, from time to time you want to see something new – a font, for instance, which is a little different and gives you an individual touch.

Is there any font that you don't like at all?

Calibri,[8] but not because it's a bad typeface, on the contrary. However, it's this particular font which currently proves that a user didn't give any thought to typography and simply applied the default font of Windows instead.

Jakob Runge was interviewed by Matthias Dufner and David Dusanek.

MEM

8 Calibri is part of a series of new typefaces introduced with Microsoft Windows Vista, but it has also been part of Microsoft Office since Windows 2007 and Mac-OS-X-Version 2008, respectively. Here it replaces Verdana as standard sans-serif type.

DRIES WIEWAUTERS

Plaque Découpée Universelle

ABCDEFGHIJKLMNO
PQRSTUVWXYZ
abcdefghijklmnopqrstuvwxyz
0123456789

PDU Regular

ABCDEFGHIJKLMNO
PQRSTUVWXYZ
abcdefghijklmnopqrstuvwxyz
0123456789

PDU Outline

FRAGILE
FRAGILE
FRAGILE
FRAGILE

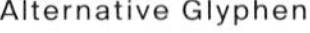

Alternative Glyphen

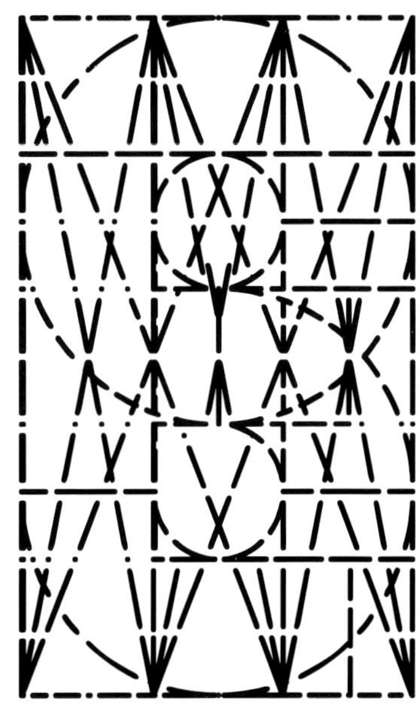

Stencil Schablone

PDU Pattern

A
B
C
D
E
F
G
H
I
J
K
L
M
N
O
P
Q
R
S
T
U
V
W
X
Y
Z

Schrift. Typeface. PDU
Gestalter. Designer. Dries Wiewauters
Label. Foundry. colophon-foundry.org
Jahr. Year. 2011

Im Jahre 1876 schuf F.A. David die Plaque Découpée Universelle, eine Schablone, mit der man jede Glyphe eines Alphabets und zusätzlich auch Muster zeichnen konnte. Mithilfe einer Reproduktion dieser Schablone wurde eine Schriftfamilie geschaffen, die viele alternative und historische Formen einschließt. Es ist ein Versuch, sowohl die unvermeidlichen Schwachstellen und Begrenzungen eines Rasters zu erforschen, als auch die unerwarteten Möglichkeiten, die sich dadurch ergeben. Die Familie benötigt viele optische Korrekturen, dennoch muss sie als die Mutter aller modularen Fonts angesehen werden.

In 1876 F. A. David created the Plaque Découpée Universelle, a stencil for drawing every glyph of an alphabet (even patterns). Using a reproduction of this stencil, a type family was constructed that contains many alternate and historic forms. It is an attempt to explore all the inherent flaws and limitations of a grid, but also the unexpected opportunities that arise therefrom. The family lacks many optical corrections, but can still be seen as the mother of modular fonts.

Der auf Print-, Typeface- und Corporate-Design spezialisierte Belgier *Dries Wiewauters*[1] (*1986) absolvierte nach dem Grafik-Design-Studium an der Kunstakademie Sint-Lucas in Gent (Belgien) seinen Master in Type Design am Werkplaats Typografie in Arnheim (Niederlanden). Zu den Inspirationsquellen des derzeit als Freelancer Arbeitenden zählen Josef Albers,[2] AutoCAD[3] und Aldo Novarese.[4] Seine wichtigen Schriften sind u.a. *PDU*, LUCA und MAD. Wiewauters wurde beim Wallpaper Graduate Directory (2011) und Power of Print (2012) ausgezeichnet.

Wann und wie wurde Schriftdesign wichtig für Sie? Warum entschieden Sie sich für einen zweiten Master-Lehrgang am Werkplaats in Arnheim? Gab es irgendwelche besonderen Gründe dafür?

dw: Ich habe schon in meinem ersten Studienjahr in Gent an Fonts »herumgekritzelt«, aber das war weiter nichts Ernsthaftes. Dann beschloss ich, mich für meinen Master an einer Schriftfamilie zu versuchen, das machte ich nebenher, denn ich entwickelte damals gerade einen Display-Font. Ich war weitgehend Autodidakt, denn die Fakultät hatte damals keinen Lehrer für Schriftdesign. Zum Glück änderte sich das, als Frederik Berlaen[5] während meiner letzten paar Monate da war. Aber statt meine Kenntnisse in Den Haag weiter zu vertiefen, entschied ich für mich, dass eine Mischung aus Type- und grafischem Design mehr mein Ding war. Und diese Möglichkeit bot sich am Werkplaats.

Gehen Sie an kommerzielle Aufträge anders heran als an persönliche Arbeiten? Fühlen Sie sich von Beschränkungen eher kreativ herausgefordert oder limitiert?

Wenn man weniger Aufträge hat, werden kommerzielle Aufträge zu persönlichen Arbeiten. Das ist beim Schriftdesign ganz praktisch, denn man hat immer etwas zu tun. Wenn man für sich selbst arbeitet, kann das allerdings beängstigend sein, denn dann neige ich dazu, noch mehr Haare zu spalten.

Beim Schriftentwurf kann die Konzeption oft ziemlich verschlungene Wege gehen. Es ist nie klar schwarz und weiß. Man kann die Vorgaben schon beim ersten Treffen mit dem Kunden abschätzen. Dann liegt es bei einem selbst, zu entscheiden, ob diese Beschränkungen eine Herausforderung sind und eine Lösung dafür zu finden. Beschränkungen können auch positiv sein, denn sie zwingen einen, eine schnelle Entscheidung zu treffen.

1 drieswiewauters.eu

2 Josef Albers (1888–1979), deutscher Maler, Bildhauer und wichtiger Bauhauslehrer. Nach der Schließung des Bauhauses durch die Nationalsozialisten musste er in die USA emigrieren und lehrte am Black Mountain College in North Carolina. Im typografischen Bereich ist Albers besonders für seine auf geometrischen Grundformen basierende Kombinationsschrift bekannt (→ S. 66).

3 AutoCAD ist eine Software vom Entwickler Autodesk, mit der technische Zeichnungen und Vektorgrafiken bearbeitet werden können. Dies ist in 2D und 3D möglich.

4 Aldo Novarese (1920–1995) war ein italienischer Typograf, der neben einer Vielzahl eigener Schriftentwürfe (u. a. Eurostile, 1962; ITC Symbol, 1984) 1956 ein Schriftenklassifizierungssystem entwickelte, das großen Anklang gefunden hat.

5 Frederik Berlaen ist ein belgischer Schriftdesigner. Nach seinem Grafik-Design-Studium am Sint Lucas in Gent, absolvierte er 2006 sein Master-Studium an der Koninklijke Academie van Beeldende Kunsten in Den Haag.

Was empfinden Sie, wenn Sie Ihre Schriften angewandt sehen?

Das ist immer spannend. Eigentlich konstruiert man lauter verschiedene Teile, die dann zu einer kleinen Maschine zusammengesetzt werden, die dann entweder dem Entwurf entspricht oder auch nicht.

Es kann auf zweierlei Art spannend sein: wenn der Designer den Font genauso gebraucht, wie man es sich vorgestellt hat; aber noch besser, wenn man von einer völlig anderen Verwendungsart überrascht wird.

Das Erste ist wie eine Bestätigung, aber das Zweite ist noch viel aufregender.

Ist Ihnen jemals ein Schriftdesigner oder eine Schrift begegnet, die Ihren Blick auf Schrift nachhaltig verändert hat?

Außer den Klassikern wie Excoffon,[6] Novarese oder Carter[7] hat mich die Arbeit von Evert Bloemsma[8] umgehauen. Er hat nur vier Familien geschaffen, doch alle vier sind anders als alles, was bis dahin geschaffen wurde, und sie alle sind voll kreativer Lösungen.

Was für ein Umfeld und welches Werkzeug brauchen Sie, um kreativ zu sein? Brauchen Sie eine besondere Umgebung?

Eigentlich nicht. Doch Reisen inspiriert natürlich sehr. Es motiviert einen, wenn man sieht, wie die Welt brummt, dann will ich auch immer einen Zahn zulegen. Aber ich brauche kein besonderes Umfeld. Solange ich genügend Ausgleich durch Laufen oder Radfahren habe, ist alles in Ordnung. Und Houjicha (gerösteter grüner Tee) und weißer Tee scheinen meine Gedanken zu beflügeln.

Sie arbeiten freiberuflich; bemühen Sie sich bewusst um Feedback von anderen? Zeigen Sie anderen Ihr »work in progress«?

Manchmal, aber die Termine geben einem nicht immer genügend Zeit dafür.

Bei den Projekten, die einen größeren Zeitrahmen haben, schicke ich regelmäßig Sachen an Leute, von denen ich weiß, dass sie einen ähnlichen Geschmack haben, und bitte sie um Feedback. Aber das sind Leute, die dieselbe Musik mögen wie ich und mit denen ich Bier trinke. Also teile ich mich ihnen wohl eher mit als dass ich Kritik von ihnen erwarte.

Ihre Schrift PDU basiert auf einer Schablone von Joseph A. David,[9] die er für Beschilderungen machte. Wie fanden Sie das Original der Schablone? Und wie fingen Sie an, es zu Ihrer eigenen Schrift zu machen? Was sind die Stärken und Schwächen der PDU?

Das geschah durch James Goggin[10] am Werkplaats, wo ich den Artikel von Eric Kindel las, den er für Typography Papers 7 schrieb. Ich war davon so fasziniert, dass ich diese Schablone unbedingt in die Hände bekommen wollte. Also zeichnete ich sie ab und ließ mir mit dem Laser drei Prototypen schneiden.

6 Roger Excoffon (1910–1983) war ein französischer Type- und Grafik-Designer, der 1947 die Werbeagentur Fonderie Olive gründete und dann half das Designbüro Studio U+O zu etablieren. Zu seinen bekanntesten Schriftentwürfen zählen Mistral (1953), eine lockere Schreibschrift, und Antique Olive (um 1960), eine Grotesk mit kalligrafischem Einschlag.

7 Matthew Carter (1937) ist ein britischer Type Designer mit umfangreicher Erfahrung mit Technologien vom Bleisatz bis zum Computer Font. Zu seinen populärsten Schriftentwürfen gehören die Bell Centennial, ITC Charter, ITC Galliard. Er wurde besonders bekannt durch die Gestaltung der frühen Web- bzw. Screenfonts Verdana und Georgia.

8 Evert Bloemsma (1958–2005) war ein holländischer Type- und Grafik-Designer, der an der ArtEZ in Arnhem und an der Academy of Art and Design (AKV) St. Joost in Breda in den Niederlanden unterrichtete. Charakteristisch für seine Schriften ist, dass diese nur aus Kurven bestehen, was wohl auf seinen Lehrer Jan Vermeulen zurückzuführen ist, der gerade Linien abgelehnt haben soll.

9 Joseph A. David ist ein amerikanischer Erfinder und Gestalter, der 1876 eine Schablone konzipierte, mit der man alle Buchstaben des lateinischen Alphabets erzeugen kann.

10 James Goggin ist ein britischer Grafik-Designer. Nach seinem Abschluss am Royal College of Art in London im Jahr 1999 gründete er das Design-Büro Practise. Ferner hat er am Werkplaats Typografie in Arnheim und an der ECAL (École cantonale d'Art de Lausanne) unterrichtet.

Und meine Faszination wurde noch größer, als ich merkte, dass man außer so ziemlich jedem Schriftzeichen des lateinischen Alphabets (manchmal mit ein wenig Mogeln, indem man die Schablone etwas nach oben oder nach unten schiebt) auch all diese exotischen Alternativen zeichnen kann. Als ich den anderen Schriftdesignern die Prototypen zeigte, wollten sie alle wissen, welche Ebay-Auktion ihnen da entgangen war. Also diente der erste Arbeitsgang erst mal dazu, all diese Leute zufriedenzustellen.

Von da ging es dann weiter, doch mir war es erst mal wichtig, diese Idee der scheinbar unendlichen Möglichkeiten einer trügerisch einfachen Schablone mit anderen zu teilen und dann zu versuchen, mir meine eigene zu machen. Schließlich versuchte ich, mit »meiner Version« einige der immanenten Stärken und Schwächen des Rasters aufzuzeigen.

Ihre Schriftentwürfe werden oft »experimentell« genannt. Sehen Sie sie auch so? Von welchem Stadium an wird eine Schrift für Sie experimentell?

Nun ja, ich betrachte sie auch als experimentell. Sie sind doch immer das Ergebnis einer bestimmten Frage: Kann dies für das benutzt werden? Was wäre, wenn ich einen Font mit X Einschränkungen entwickle, was wäre, wenn …

Es fängt immer mit einer bestimmten Frage an, die ich mir stelle, was aber nicht bedeutet, dass man am Ende auf diesen Anfang schließen können sollte. Das Experiment ist immer nur der Impuls, der einen Designprozess in Gang setzt. Selbst bei einer Auftragsarbeit sehe ich es als ein Spielen mit den vorgegebenen Einschränkungen an.

Ist Ihnen Lesbarkeit wichtig? Denken Sie daran, ehe Sie mit der Arbeit anfangen?

Ja, natürlich. Man baut ja auch kein Auto, das nicht fahren kann. Dasselbe gilt für Fonts. Man kann die Geschwindigkeit eines Autos festlegen, aber es sollte sich doch immer irgendwie bewegen.

Entwickeln Sie lieber Textschriften oder Schriften für Display, und aus welchem Grund?

Mit Display-Schriften ist man schneller fertig, aber Fonts, mit denen man auch Texte setzen kann, sind intellektuell lohnender. Aber für eine vollständige Schriftenfamilie mit sämtlichen Gewichten hatte ich bisher weder genügend Geduld noch genügend Geschick.

Arbeiten Sie lieber analog oder digital, und aus welchem Grund?

Das hängt ganz vom jeweiligen Projekt ab. Oft zeichne ich erst alles auf Papier, mit verschiedenen Mitteln, aber dann kommt schon die digitale Arbeit dazu.

Obwohl Type Design überwiegend ein digitaler Prozess ist, sollte man, wenn man Zweifel hat, doch lieber schnell zeichnen, um es auszuprobieren, ehe man damit an den Bildschirm geht. Einen stilistischen OpenType-Satz zu programmieren, kann perfekt geeignet sein, um verschiedene Zeichnungen auszuprobieren.

Wie fangen Sie an, wenn Sie eine neue Schrift entwerfen? Gibt es da gewisse Rituale oder Gewohnheiten, die Sie im Laufe der Zeit entwickelt haben?

Eigentlich nicht. Wenn ich mit der digitalen Arbeit anfange, fange ich immer damit an, das Gewicht für die beabsichtigte Größe festzulegen, denn ich finde, die Schwärzung eines Fonts ist eines seiner wichtigsten Merkmale. Aber abgesehen davon, habe ich keine festgelegten Arbeitsweisen.

Was ist für Sie das Reizvolle an der Ungewissheit oder am Zufall beim Designprozess? Denken Sie darüber nach, ehe Sie anfangen, oder ergibt es sich ganz von allein?

Zufälle sollte man nicht planen, das würde nur zu einer falschen Wahllosigkeit führen. Wenn sie sich jedoch während des Prozesses ergeben, dann sollte man sie immer ernst nehmen. Selbst wenn sich herausstellen sollte, dass etwas doch nicht funktioniert, ist es immer noch lohnender, sich mit dem »Warum« zu befassen, als einfach zu akzeptieren, dass es nicht in das Gesamtbild des Designs passt.

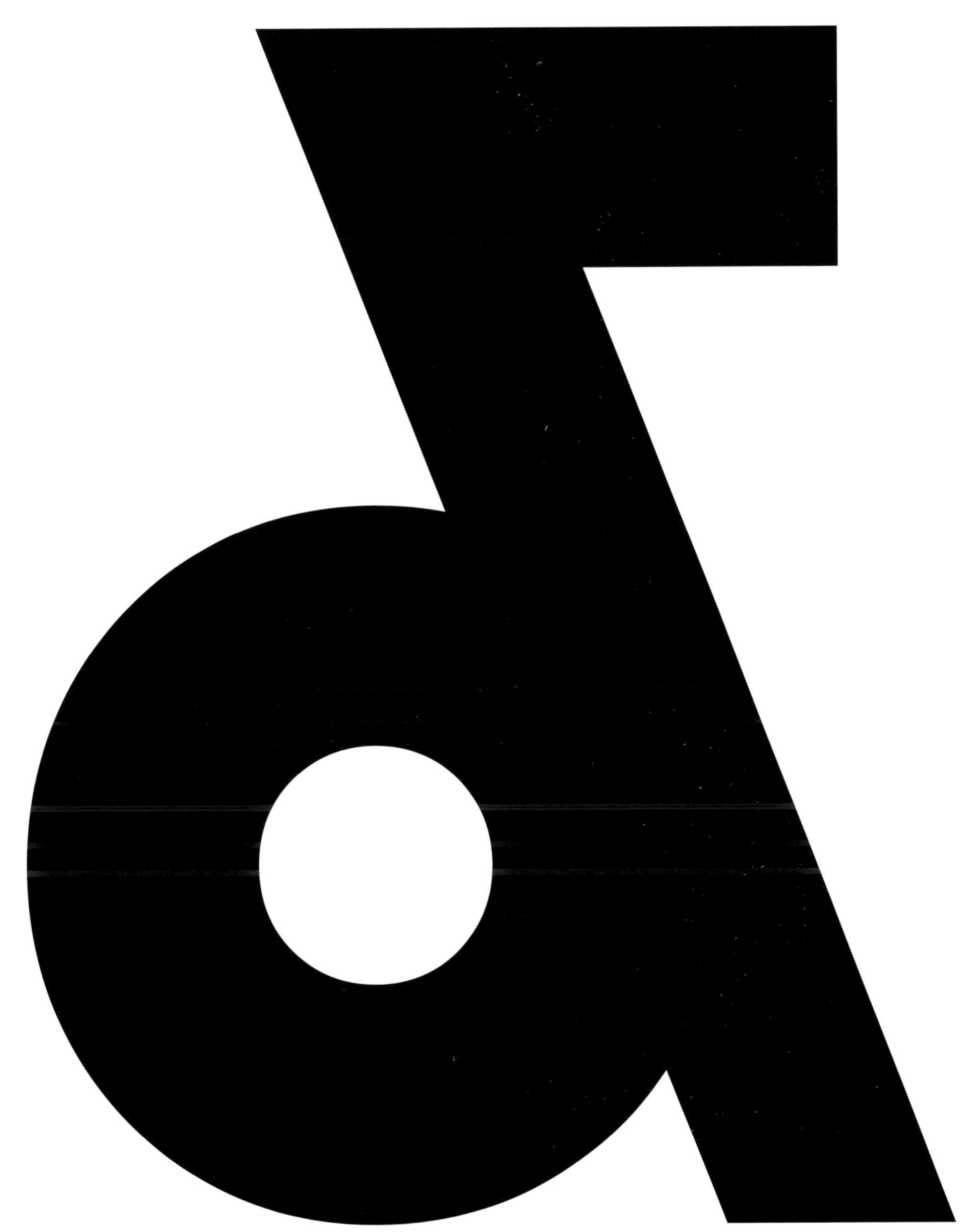

Warum basiert ein Teil Ihrer Arbeit auf historischen Vorlagen? Nach welchen Kriterien wählen Sie die Autoren dieser Vorlagen aus? Und tun Sie etwas Bestimmtes, um dafür zu sorgen, dass es »Ihr Werk« ist und eben nicht nur eine Kopie?

Wenn ein historisches Vorbild mich anregt und ich das Gefühl habe, dass es im Laufe der Zeit verlorengegangen ist, oder dass es ein Problem aufzeigt, das man vielleicht auf andere Art und Weise lösen kann, dann probiere ich gern aus, ob es noch immer gültig ist.

Ich glaube, Schriftdesign ist noch mehr wie ein Ouroboros als jede andere ästhetische Disziplin. Daran sollte man immer denken und versuchen, eher der Kopf der Schlange zu sein als die Schwanzspitze.

Wie wissen Sie, wann eine Schrift fertig ist? Verlassen Sie sich da auf Ihr Bauchgefühl, oder suchen Sie dazu auch andere Meinungen?

Eine Schrift ist niemals ganz fertig. Sie ist lediglich ein Abbild des Entwicklungsstadiums zum Zeitpunkt der Abgabefrist.

Sind Ihnen Trends wichtig? Versuchen Sie, den »Zeitgeist« in Ihrer Arbeit festzuhalten? Wo finden Sie Ihre Inspiration?

Trends sind mir nicht wichtig. Und einen Zeitgeist gibt es immer, auch wenn man eine Schrift entwickeln würde, ohne ständig von der globalen visualisierten Kultur im Netz bombardiert zu werden. Mich inspirieren Orte, an denen die Tradition erhalten geblieben ist. Solange man nicht ausdrücklich danach sucht, kann man in allem und an jedem Ort Inspiration finden.

Gibt es beim Schriftdesign etwas, das Ihnen unsympathisch ist, wie zum Beispiel die Zurichtung?

Und wie! Alles, was über die rein ästhetische Arbeit hinausgeht, dauert mir viel zu lange. Allerdings geben einem diese mönchischen Momente auch Gelegenheit, um darüber nachzudenken, ob man wirklich alle Probleme gut gelöst hat.

Glauben Sie, dass Schriftdesigner eine besondere Begabung brauchen? Haben Sie Ratschläge für Leute, die Schriftdesigner werden wollen?

Nehmt euch viel Zeit, um Ideen auszuprobieren und euch Sachen anzusehen, die bereits existieren, und tut das, was ihr gefunden habt, in einen Topf. Kümmert euch nicht um die ästhetischen Elemente, denkt aber darüber nach, warum diese Elemente für eine bestimmte Schrift zwingend sind, um ein Gesamtbild entstehen zu lassen.

Das Interview mit Dries Wiewauters führten Alexa Spors und Lisa Steinhauer.

Dries Wiewauters[1] (*1986), a Belgian designer specialized in print, typeface and corporate design, graduated at the Sint-Lucas School of Arts, Ghent, and got his Master Degree in type design at Werkplaats Typografie in Arnhem, Netherlands. He is working freelance and is inspired by Josef Albers,[2] AutoCAD[3] and Aldo Novarese.[4] His most important typefaces are, among others, *PDU*, LUCA and MAD. Wiewauters was honoured by the Wallpaper Graduate Directory (2011) and by Power of Print (2012).

When and how did type design become something essential for you? Why did you decide to get a second Master's degree at Werkplaats in Arnhem? Were there any special factors that led to this decision?

dw: I was already doodling fonts in my first year as a BA in Ghent, but nothing serious. Then for my MA, I decided to have a go at a text family, so I did that besides some other display font. I was mostly self-taught since the department did not yet have a teacher in their ranks, who could teach type design. That luckily changed when Frederik Berlaen[5] got involved in the program during the last few months I was there. But instead of fully focussing on type design, I decided that the mix between both type and graphic design was more my thing. It turned out that this was possible at the Werkplaats.

Do you approach commercial work in a different way than personal work? Do restraints fuel your creativity or do they limit the way you act?

When there are fewer jobs, commercial work becomes personal work. This comes in handy in type design since you will always have something productive to do.

Working for yourself can be daunting though, because then I tend to be even more nit picky.

For drawing type the approach is often convoluted. It is not clearly black and white. You can often judge the constraints through the first contact with the client. It is then up to yourself to decide if these are constraints in which you find it interesting to arrive at a solution. Constraints can be rewarding though, since they forcefully speed up the decision-making process.

What do you feel when you see your typefaces used in projects?

It is always exciting. You are basically constructing all these parts that can be combined into small machines that can either work for or against somebody's design.

The most exciting use can be both – when the designer uses the font for the qualities you tried to put into it or, even better, when you are surprised by a completely different way of use.

The first can be taken as a thumbs-up of sorts, the latter, however, is even more exciting.

Did you ever come across a special type designer or typeface that permanently changed the way you see type now?

Besides the classics from the likes of Excoffon,[6] Novarese, Carter,[7] the work of Evert Bloemsma[8] is mind-blowing. He only got to make four families, yet they are all different from the prior typographic history and they all contain creative solutions.

What do you need to be creative in terms of environment and tools? Do you need special surroundings?

Not really. Travelling inspires you more. It is motivating to see the pace at which the world buzzes, it makes me want to crank up the pace. Certain surroundings are not a necessity though. In general, as long as I get enough balance by going for a run or cycling all is well. Houjicha (roasted green tea) and white tea seem to help to get my mind in the right place, though.

Since you are working as a freelancer, do you consciously seek feedback from others? Do you share your »work in progress«?

Sometimes, but often deadlines do not always allow any extra space for this.

For the more long term projects I regularly send out stuff to people, who have similar aesthetics, to get general feedback. However these are also people I share music and beer with. So it is more a pool of »sharing« than asking for critique I guess.

1 drieswiewauters.eu

2 Josef Albers (1888–1979), German painter, sculptor and important Bauhaus teacher. After the Bauhaus was closed by the Nazis he emigrated to USA, where he taught at the Black Mountain College in North Carolina. In typography, Albers is well-known for his combination type, based upon geometric principles (→ S. 69).

3 AutoCAD is a software by the developer Autodesk, suitable for working on technical drawings and vector graphics, both in 2D or 3D.

4 Aldo Novarese (1920–1995) was an Italian typographer, who apart from numerous type designs (e.g. Eurostile, 1962; ITC Symbol, 1984) developed a system of type classification which became very popular.

5 Frederik Berlaen is a Belgian type designer. After studying Graphic Design at Sint Lucas in Ghent, he got his Master Degree at the Koninklijke Academie van Beeldende Kunsten in Den Haag.

6 Roger Excoffon (1910–1983) was a French type and graphic designer who in 1947 launched the Marketing Agency Fonderie Olive and then helped to establish the design studio U+O. Among his best-known type faces are Mistral (1953), an unconstrained script, and antique Olive (around 1960), a Grotesque with calligraphic elements.

7 Matthew Carter (*1937) is a British typographer and type designer with experience of typographic technologies ranging from hand-cut punches to computer fonts. Among his popular typefaces are Bell Centennial, ITC Charter, ITC Galliard. He became particularly known for the design of the early web- and screenfonts Verdana and Georgia.

8 Evert Bloemsma (1958–2005) was a Dutch type designer and graphic artist who taught at the ArtEZ in Arnhem and at the Academy of Art and Design (AKV) St. Joost in Breda, Netherlands. Characteristically, his typefaces consist mostly of curves, which is thought to go back to his teacher Jan Vermeulen, who rejected straight lines.

»CONSTRAINTS CAN BE REWAR-DING, SINCE THEY FORCEFULLY SPEED UP THE DECISION-MAKING PROCESS.«

Dries Wiewauters
→ S. 173

Your typeface PDU is based on a stencil by Joseph A. David[9] and was made for signs. How did you find the original stencil? How did you begin the process of transforming it into your own? What are the strengths and weaknesses of PDU?

It was through James Goggin[10] at the Werkplaats that I got into the article by Eric Kindel he had written for Typography Papers 7. I got so fascinated that I felt the need to have the object in my hands. I drew it out and got three prototypes lasercut.

The fascination only grew because it turned out that besides every letter of the Latin alphabet (sometimes with minor cheating by shifting the stencil up or down), you could also do all these funky alternative characters. When I showed the prototypes to other designers they were all wondering which Ebay auction they had missed. So a first run was in order to get all these people their fix.

It all went from there, however for me it is more about sharing this idea of the seemingly limitless possibilities of a deceptively simple stencil, then trying to make it my own. In the end I tried to make »my version« highlight some of the grid's inherent strengths and weaknesses.

Your typefaces are often called »experimental«. Are they experimental for you? At what point does a typeface become experimental to you?

Well, I also consider them experimental. They are always the result of a certain question: can this be applied to that, what if you design a font with X constraints, what if …

They always start with a certain question I have, but this does not mean the starting point should be obvious in the end. The experiment is always this impetus that kickstarts a design. Even for a commercial job I see it as a game that you play within the given constraints.

Is legibility an important aspect for you? Is it something you think about before you start working?

Yes. Why would you design a car that does not move? The same goes for fonts. You can set the speed the car needs to be able to attain, however it should always move somehow.

Do you prefer to design text typefaces or display typefaces? If so, why?

Display typefaces are a quicker »fix«, however fonts that can also be used to set text are intellectually even more rewarding. But finishing an entire multi weight family is something for which I have – so far – neither got the patience or the dexterity.

What do you enjoy doing more: Working analog or digitally? Why?

It depends on the project. Often I sketch first on paper with different utensils, which then gets intertwined with digital work.

Although it is mostly a digital process, when in doubt it is always better to quickly fall back on sketching to test something out before taking it to the screen. Programming a stylistic OpenType set can be the perfect way to test different sketches.

How do you start designing a new typeface? Do you have any rituals or routines that you have developed over the years?

Not really. When the process gets digital, I always start by setting the weight to match the intended sizes first. I think the blackness of a font is one of the most important aspects. But apart from this routine I do not really have a fixed production schedule.

What is attractive about using uncertainty or coincidence in your process of designing typefaces? Do you think about it before you even start or is it something that comes naturally to you?

Coincidences should not be planned, that would just result in fake randomization. However, if they occur during the process, I think they should always have a place. Even if it turns out that a certain thing does not work, figuring out why it does not work is more rewarding than knowing that something does not fit into the overall aesthetics of a design.

Why do you base some of your work on historic material? How do you choose the original authors of said material? Is there anything in particular that you do to ensure that it is still »yours« and not just a basic copy?

If something historic triggers me and I feel it has been lost over time or it has some problems to offer that could perhaps be solved differently, I try to see if it still works.

Maybe even more than other aesthetic fields, type design is like an ouroboros. I feel it is better to be aware of this and try to be the head of the snake instead of the tail.

How do you know that you are done with designing a typeface? Do you rely on your own gut feeling or do you invite feedback from external sources?

A typeface is never done. It is just a picture of the state it was in when the deadline came.

Are trends important to you? Do you try to incorporate »Zeitgeist« into your work? How do you find inspiration?

Not really. Zeitgeist will always be there, even if you were to design fonts without ever getting bombarded through the globalized visual culture in the net. I always get inspired when I find local places fixed in tradition. As long as you do not try to search for it, inspiration can be found everywhere and in everything.

Is there anything you dislike about designing typefaces, for example kerning?

Oh yes! Everything besides figuring out the aesthetic hurdles is taking too long. However, the sometimes monk-like moments are also a state where you can examine whether you have found the answers for these hurdles.

Do you think type designers need a special set of skills? Do you have any advice for people who want to start designing typefaces?

Spend a lot of time trying things out, and a lot of time researching existing stuff, mix up what you've found, but do not look for aesthetic elements. Look for the reason why those elements can be considered mandatory for the system that results in the overall aesthetics.

Dries Wiewauters was interviewed by Alexa Spors and Lisa Steinhauer.

9 Joseph A. David is an American inventor and designer who in 1876 created a stencil for drawing all the letters of the Latin alphabet.

10 James Goggin is a British graphic designer. After graduating in 1999 at the Royal College of Art in London he started the design studio Practise. He has also taught at the Werkplaats Typografie in Arnhem and at the ECAL (École cantonale d'Art de Lausanne).

Zweck einer Rundung ist, Platz für die Darstellung zu sparen, insbesondere bei Gleitkommazahlen und Dezimalbrüchen.

Roletta Sans ExtraBold – 24pt

Die Genauigkeit des Ergebnisses ist der darstellbaren bzw. messbaren Einheit anzupassen. Also die kleinste mögliche Währungseinheit wie z.B. Cent oder ganze Gramm bei Küchenwaagen.

Roletta Slab Bold Italic – 16pt

Doch werden auch große Ganzzahlen gerundet. Zum Beispiel rundet die Bundesagentur für Arbeit die errechnete Anzahl der Arbeitslosen auf volle 100.

Roletta Slab Medium – 10pt

Wird eine positive Zahl vergrößert, so spricht man von Aufrunden, wird sie verkleinert, von Abrunden. Bei negativen Zahlen sind diese Wörter doppeldeutig. Werden Nachkommastellen nur weggelassen, spricht man von Abschneiden. Das Runden verändert in den meisten Fällen den Wert der gerundeten Zahl.

Roletta Sans Regular – 6pt

Sans Regular, *Italic*
Sans Medium, *Italic*
Sans Bold, *Italic*
Sans ExtraBold, *Italic*
Sans Black, *Italic*

Slab Regular, *Italic*
Slab Medium, *Italic*
Slab Bold, *Italic*
Slab ExtraBold, *Italic*
Slab Black, *Italic*

Schrift. Typeface.	Roletta
Gestalter. Designer.	Andrea Tinnes
Label. Foundry.	typecuts.com, primetype.com
Jahr. Year.	2004–2010

PTL Roletta ist eine umfangreiche runde Schriftfamilie, die aus einer Sans Serif/Grotesk und einer serifenbetonten Linear-Antiqua besteht und in der Funktionalität auf Verspieltheit trifft. Während die Rundungen von Roletta bei kleinen Größen zurückhaltend und kaum zu sehen sind, geben sie der Schrift doch in Displaygröße einen ganz eigenen Charakter mit vielen Besonderheiten.

Mit fünf Gewichten pro Stil, einschließlich Kapitälchen und Kursiver, sind die Roletta Sans und Slab ausgezeichnete Gefährten und gut geeignet für viele Textanwendungen.

Die OpenType-Version beinhaltet viele fortschrittliche Features mit verschiedenen Zahlensätzen, alternativen Zeichen und geometrischen Symbolen.

PTL Roletta umfasst auch sechs verschiedene ornamentale Fonts für endlose Möglichkeiten dekorativer Anwendungen und Muster. Alle ornamentalen Fonts befassen sich mit der Rundung: entweder gepunktete Ornamente, runde Formen und Umrisse oder runde Blütenformen.

PTL Roletta is a comprehensive round family consisting of a sans and a slab variant, combining functionality and playfulness. While Roletta's round shapes are discrete and barely visible at small sizes, they give the typeface its distinctive character with many refined details when used at display sizes.

With five weights per style, including caps and italics, PTL Roletta Sans and Slab are excellent companions, well-suited for a wide range of text uses.

The OpenType version includes many advanced layout features with various figure sets, alternative characters and geometric symbols.

PTL Roletta also comes with a set of six ornament fonts, offering an endless range of decorative composites and patterns. All ornament fonts play with the notion of roundness: either dotted ornaments, circular forms and outlines, or roundish floral shapes.

RESULTAT

~~*Logarithmus*~~

Kommastelle

Dezimalbruch

-Aufrunden-

SYSTEMATIK

Roletta Slab Bold
Roletta Sans Italic
Roletta Sans Regular
Roletta Sans ExtraBold Italic
Roletta Sans Medium Italic
Roletta Slab Black

Die gebürtige Saarländerin *Andrea Tinnes*[1] (*1969) studierte Kommunikationsdesign an der FH Mainz und absolvierte den Master of Fine Arts für Grafik-Design am California Institute of the Arts. 2004 gründete sie in Berlin das Fontlabel »Typecuts«.[2] Von 2003 bis 2008 lehrte Andrea Tinnes Schrift und Typografie an der Bergen National Academy of the Arts in Norwegen und seit 2007 an der Burg Giebichenstein Kunsthochschule Halle. Ihr Schriften-Spektrum reicht von der ornamentalen Volvox[3] über experimentelle Fonts bis hin zu Schriftsystemen wie der Skopex.[4] Ihre Arbeiten wurden unter anderem in »Graphic Design for the 21st Century« und im »Eye Magazine« publiziert und mehrfach ausgezeichnet (TDC53, Red Dot).

Wie manifestiert sich in Deinen Augen die Typo-Szene in Berlin?

at: Mittlerweile hat sich in Berlin eine sehr lebendige und große Type Designer-Szene etabliert und es findet ein regelmäßiger Austausch unter den Gestaltern statt. Die Behauptung von Erik Spiekermann,[5] dass in den »Niederlanden fast auf jeden Quadratmeter ein Type Designer kommt«, trifft sicherlich noch immer zu. Mittlerweile lassen sich jedoch immer mehr Schriftgestalter nach ihrem Abschluss in Den Haag[6] oder Reading[7] in Berlin nieder. Dadurch entsteht vielleicht eine gewisse Konkurrenz, aber ich finde es großartig, in einer Stadt zu leben, in der dieser Austausch tatsächlich stattfindet. Es gibt viele Freundschaften unter Type Designern. Manche haben ihren Master zusammen gemacht, sich auf Konferenzen getroffen oder sich auf privater Ebene vernetzt. Ich habe z.B. durch Verena Gerlach[8] einige spannende Gestalter kennengelernt. Durch das gemeinsame Interesse kommt man zusammen und tauscht sich aus. Man trifft sich zum Abendessen oder auf diversen Events.

Wie blickst Du auf Deine Zeit in Mainz und auf Dein Studium zurück?

Diese Frage ist wahrscheinlich etwas heikel, aber ich blicke sehr kritisch auf Mainz zurück. Wobei ich betonen möchte, dass es zu meiner Zeit eine ganz andere Hochschule war. Die damalige Situation lässt sich in keiner Weise mit

1 typecuts.com

2 Typecuts ist ein 2004 von Andrea Tinnes gegründetes unabhängiges Schrift- und Grafik-Label, spezialisiert auf die Kreation und Produktion zeitgenössischer Schriften.

3 Volvox ist ein aus fünf Schnitten bestehendes Schriftsystem von Andrea Tinnes, das durch die Ästhetik wissenschaftlicher Zeichnungen inspiriert wurde. Durch Überlagerungen ermöglicht es den Entwurf von komplexen Mustern und Strukturen.

4 Skopex ist eine von Andrea Tinnes gestaltete serifenlose Schriftfamilie mit vertikaler Ausrichtung, die 2005 bei Primetype veröffentlicht wurde. Merkmale der PTL Skopex Gothic sind die schrägen Strichendungen der Ober- und Unterlängen.

5 Der international renommierte Grafiker, Typograf und Schriftgestalter Erik Spiekermann (*1947) gründete 1979 das Designbüro MetaDesign in Berlin später mit Filialen in San Francisco und Zürich. Er ist Mitbegründer der internationalen Vertriebsplattform FontShop und vertreibt seine Schriften zusammen mit Joan Spiekermann und Neville Brody unter FontFont. Eine seiner bekanntesten und einflussreichsten Schriften ist die FF Meta.

6 Die Koninklijke Academie van Beeldende Künsten, Den Haag ist eine 1682 gegründete Kunsthochschule in den Niederlanden. Seit 2002 wird hier der Masterstudiengang Type and Media angeboten.

7 Die University of Reading ist eine Universität im englischen Reading. Sie wurde 1926 gegründet, besonders einflussreich ist der Masterstudiengang Typeface Design. Viele der in diesem Buch genannten Schriftgestalter haben dort studiert bzw. gelehrt (→ S. 58).

8 Verena Gerlach (→ S. 124)

der heutigen vergleichen. Ich glaube, dass fast das komplette Kollegium gewechselt hat.

Als ich Ende der achtziger Jahre mit meinem Studium in Mainz angefangen habe, hatte der Computer gerade seinen Siegeszug begonnen, die digitale Revolution und die Postmoderne waren im Bereich Grafik-Design in aller Munde. Emigre, Eye Magazine, später Ray Gun, Carson[9] und Typography Now, damals die Bibel für viele Studierende, hatten einen großen Einfluss. Dieses hat sich aber in keiner Weise in der Lehre oder im Diskurs an der FH widergespiegelt. So entstand eine gewisse Frustration unter den Studierenden. Für mich hat der theoretische, vielleicht auch philosophische Diskurs in der sehr praxisorientierten Lehre gefehlt, die Inspiration von Seiten der Lehrenden. Das war ein Grund dafür, mich nach CalArts in den USA zu orientieren.

Allerdings will ich auch festhalten, dass ich unter Hans Peter Willberg[10] sehr viel gelernt habe. Buchgestaltung war damals mein Schwerpunkt, und so habe ich von ihm das A und O der guten Typografie gelernt. Natürlich schon mit dem Blick eines Fachmannes, der aus der klassischen Typografie / Buchgestaltung kommt und zunächst vielleicht kein allzu großes Interesse an neuen und spannenden Gestaltungstendenzen in anderen Ländern hatte. Ich muss ihm aber zu Gute halten, dass er zum Ende seiner Lehrzeit unglaublich offen und interessiert an experimenteller Gestaltung in den Ebenen Grafik-Design, Typografie und Schrift war. Das finde ich beeindruckend und würde es für mich gerne zum Vorbild nehmen, dass man immer offen und neugierig bleiben kann, gerade als Lehrende, und in seiner Meinung und Haltung nie festgefahren ist.

Hast Du in Kalifornien diese Offenheit und den Diskurs mit der zeitgemäßen Gestaltung gefunden? Und konntest Du die klassische, akkurate Gestaltung aus Deutschland dort für Dich nutzen?

Man wirft CalArts bzw. den Lehrenden und Studierenden im Studiengang »Graphic Design« oft vor, dass sie sehr »stilistisch« gestalten, dass sie sich nur auf die Form und die visuelle Sprache konzentrieren. Aber tatsächlich findet der Unterricht der Designtheorie und -geschichte dort auf einem hohen Niveau statt. Sowohl theoretisch als auch praktisch, historisch und zeitgenössisch tauscht man sich über alle Facetten der Typografie und des Grafik-Designs aus und reflektiert die Disziplin. Ich konnte sehr viel an Erfahrungen und Eindrücken mitnehmen und habe erst dort begonnen, mich mit Designgeschichte auseinanderzusetzen. Ich habe in dieser Zeit viele spannende Persönlichkeiten mit ihrem gestalterischen und künstlerischen Werk kennengelernt, die mir bis dahin nicht vertraut waren.

Mein ursprünglicher gestalterischer Ansatz war akkurat und akribisch, sehr auf das Detail fixiert. Diese Form von »Gründlichkeit« in der Gestaltung habe ich mitgenommen. Natürlich in der Hoffnung, dass ich mich nach anderen Positionen umschauen, mich weiterbilden und gestalterisch öffnen kann, um einen anderen Blick und eine andere Haltung zu entwickeln. Allerdings war es genau diese deutsche Akkuratesse, an der die kalifornischen Lehrenden interessiert waren. Wie man diese sehr klassische Haltung, das solide Handwerk der Ausbildung bei einer Koryphäe wie Willberg, mit einem sehr freien und malerischen Umgang mit Schrift und Typografie kombinieren kann, war auf jeden Fall ein wichtiges Thema. Mein formaler gestalterischer Ansatz ist an CalArts experimenteller und vielleicht auch organischer geworden. So konnte ich meinen eigenen Stil entwickeln: Auf der einen Seite ist er sehr spielfreudig, sehr am

9 Der US-amerikanische Typograf, Grafiker und Surfer David Carson (*1957) wurde bekannt für seine innovative Magazingestaltung und die Verwendung experimenteller Typografie. Sein Typografie- und Layout-Stil prägte die Grunge-Ästhetik. Carson war Art Director der Zeitschrift Ray Gun, bevor er 1995 sein eigenes Studio in New York gründete (→ S. 72).

10 Hans Peter Willberg (1930–2003) war ein deutscher Typograf, Illustrator, Buchgestalter, Hochschullehrer und Fachautor. Er war Geschäftsführer der Stiftung Buchkunst in Frankfurt am Main und 1975 bis 1996 Professor für Typografie und Buchgestaltung an der FH Mainz. Zu seinen bedeutendsten Werken zählen »Lesetypografie«, »Erste Hilfe in Typografie« sowie »Wegweiser Schrift«.

Detail interessiert, anderseits akkurat und fast schon perfektionistisch.

Deine Fonts sind sehr experimentell, aber auch sehr gut ausgebaut. Was müssen Fonts für Dich persönlich leisten?

Das kommt immer auf den Kontext an. Es gibt ganz unterschiedliche Ansätze Schriften zu gestalten: die spielerische oder experimentelle Auseinandersetzung mit einem inhaltlichen Konzept, die Gestaltung von plakativen Alphabeten, die visuell ansprechend sein sollen, oder die Entwicklung von Textschriften für den Gebrauch. Für diese drei Kategorien experimentell, plakativ und funktional, gibt es selbstverständlich ganz unterschiedliche Ansprüche.

Was Textschriften betrifft, so ist es für mich wichtig, dass sie einen guten Rhythmus haben, und, was wahrscheinlich die Herausforderung schlechthin für jeden Type Designer ist, dass sie etwas Eigenes haben. Sie müssen originell sein, ohne ihre Eigenheiten zu sehr in den Vordergrund zu stellen. Sonst besteht die Gefahr, dass man schnell müde wird, sich die Schrift anzuschauen oder sie zu benutzen. Für den Gebrauch sollte die Schrift so ausgebaut sein, dass man einen Fundus an Möglichkeiten hat, wie Ligaturen, alternative Zeichen, unterschiedliche Ziffernsätze – also all das, was über OpenType heutzutage möglich ist. Damit Typografen auch wirklich gut damit arbeiten können, war das Ziel meiner letzten größeren Schriftfamilien immer, einen gut ausgebauten Zeichensatz zu bieten.

Kannst Du uns eine kurze Lebensgeschichte der Roletta zusammenfassen?

Diese Schrift hat tatsächlich eine längere Lebensgeschichte, weil die Entwicklung sich über viele Jahre hingezogen hat. Die ursprüngliche Idee entstand im Rahmen eines Pitches für den Rundfunk Berlin-Brandenburg. Wir wollten eine Wortmarke bzw. eine Hausschrift schaffen, welche die Kontraste, die Vielseitigkeit, sowie die Fusion der beiden Bundesländer aufgreift. Die postmoderne Idee, verschiedene Charakteristika in der Schrift zu vereinen, zeigte sich durch die Kombination aus runden und eckigen Abschlüssen.

Der weitere Anspruch war, eine serifenlose Schrift zu gestalten, die einerseits platzsparend mit ihren Versalien auf dem Bildschirm angewendet werden kann und die andererseits in ihren Kleinbuchstaben einen etwas lebendigeren Rhythmus aufzeigt, so dass sie gut differenzierbar und lesbar ist. Mit diesem Vorsatz habe ich begonnen, die damalige RBB-Schrift zu entwickeln. Es gab sie zunächst rudimentär in vier Gewichten. Den Pitch konnten wir damals leider nicht für uns entscheiden, und so hat die Schrift einige Jahre geruht, bis ich das Design wieder aufgegriffen habe.

Der postmoderne Ansatz der RBB-Schrift mit ihren runden und eckigen Abschlüssen erinnerte mich letztendlich dann doch zu sehr an die 90er Jahre. Ich musste mich also entscheiden, ob die Formen komplett gerade oder rund werden. Nachdem ich den Zeichensatz zunächst begradigt hatte, habe ich mich mit Ole Schäfer[11] getroffen und wir mussten beide feststellen, dass der andere Weg – komplett rund zu werden – wesentlich interessanter ist.

Im Jahr 2005 habe ich dann begonnen zu recherchieren, wie viele runde Schriften es auf dem Markt gibt, z.B. die Sauna[12] von Underware[13] oder die Hermes[14] von Optimo.[15]

11 Ole Schäfer (* 1970) ist ein deutscher typografischer Gestalter und Schriftentwerfer. Er war involviert in den Ausbau der ITC-Schrift Officina, der FF Meta sowie der FF Info. 2000 veröffentlichte er seine FF Fago, zwei Jahre später gründete er den Schriftenverlag Primetype (→ S. 125).

12 Die Sauna ist eine serifenlose Schriftfamilie, gestaltet von Bas Jacobs, Akiem Helmling und Sami Kortemäki. Sie wurde 2002 bei Underware veröffentlicht und besteht aus 18 Schnitten in drei Gewichten.

13 Underware ist eine Type Foundry und ein Studio für Grafik-Design, gegründet 1999 von Akiem Helmling, Bas Jacobs und Sami Kortemäki, das nicht nur Fonts produziert, sondern mit Typeradio auch einen Radiosender zum Thema Schrift gründete.

14 Hermes ist eine Neuinterpretation der in der Hermes 3000 eingesetzten Schreibmaschinenschrift. Durch einfache geometrische Formen und abgerundete Enden verkörpert die Schrift einen warmen und freundlichen Modernismus. Sie wurde 2002 vom Grafik-Design-Studio Gavillet & Rust bei Optimo veröffentlicht.

15 Optimo ist eine unabhängige Type Foundry aus Genf (CH), gegründet 1998 von Gilles Gavillet und David Rust. Sie entwickelt und vertreibt qualitativ hochwertige Schriften, u. a. Schriftfamilien wie Theinhardt, Hermes, DidotElder, Dada Grotesk und Stanley.

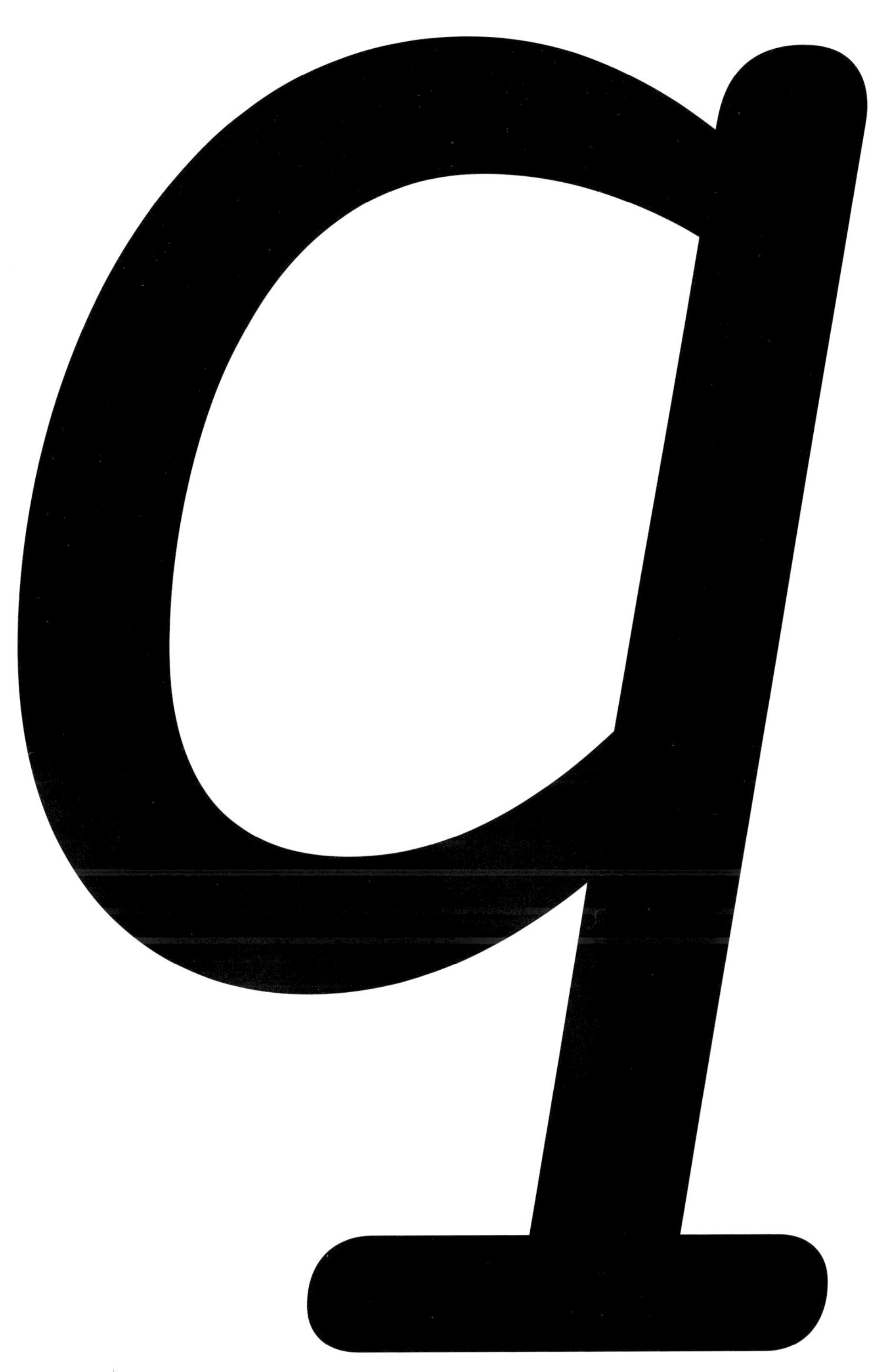

Floral L
-
Floral l
+
OutlineDots U
+
Floral s
=
Dots q
+
Dots Q
=
Floral v
+
Floral K
+
OutlineDots o
=
Dots L
+
Outline a
-
OutlineBackground S
+
OutlineDots S
=
Floral a
+
Floral A
=
Circular X
-
OutlineBackground s
+
OutlineDots s
=

Roletta Ornaments

Ich habe mir weiterhin die Fleischmann,[16] die Cooper Black[17] und die Helvetica Rounded[18] angeschaut, um auszuloten, wie ich mich mit einer runden Schrift jenseits dieser Schriften positionieren kann. Tatsächlich gab es zu jener Zeit keine umfangreiche Schriftfamilie, die Sans und Serif miteinander verbindet. Das war die Initialzündung für die Idee, eine umfangreiche runde Familie zu entwickeln.

Durch zahlreiche Projekte und berufliche Veränderungen verlängerte sich die gesamte Entwicklung der Roletta enorm. Als sie dann 2010 endlich fertig war, war der Boom der runden Schriften (der tatsächlich zwischen 2006 und 2010 stattfand) fast schon vorbei. Nichtsdestotrotz war diese Schrift für mich persönlich ein wichtiger Entwicklungsschritt, weil ich mir viele formale Grundlagen über das Schriftenmachen aneignen konnte. Und es hört an dieser Stelle noch immer nicht auf, denn jetzt wird die Geschichte der Roletta als »unrunde, begradigte« Burg Grotesk weiter geführt.

Was macht die Roletta zum Hingucker und was sind ihre spezifischen Formen?

Das sind wahrscheinlich der Kontrast und die Diskrepanz zwischen kleinen und großen Punktgrößen. In kleinen Größen nimmt man sie gar nicht als runde Schrift wahr, sondern als zurückhaltende, serifenlose Textschrift. Je größer die Schrift wird, desto mehr sieht man die einzelnen Details, sie wird weicher, freundlicher und verspielter. Das besondere Merkmal ist, dass die Versalien in der Proportion eher schmal sind und im Verhältnis zu den Kleinbuchstaben eher konstruiert wirken. Das heißt, wenn man nur in Versalien setzt, ist die Roletta eine sehr modular durchkonstruierte, serifenlose Schrift. Die Kleinbuchstaben hingegen sind in ihrem Duktus wesentlich verspielter und lebendiger. Sie ergeben auch im Textbild einen anderen Rhythmus. Die Schrift schafft somit einen breiten Spielraum in der Gestaltung.

Das Interview mit Andrea Tinnes führten Yvonne Kümmel und Jens Giesel.

16 Die Schriftfamilie DTL Fleischmann geht auf die historischen Schriftschnitte von Johann Michael Fleischmann zurück und besteht aus Text- sowie Display-Schnitten. Der Leipziger Designer Erhard Kaiser orientierte sich stark am original Bleisatz-Schriftbild und veröffentlichte die Antiqua-Schriften 1997 bei der Dutch Type Library.

17 Oswald Bruce Cooper entwarf 1921 für ein Plakat eine fette Schrift mit rundlichen Serifen, die er 1922 bei Barnhart Brothers & Spindler herausbrachte. Die Cooper Black entsprach dem damaligen Werbe-Zeitgeist: einfach, freundlich, kräftig.

18 Die Schriftfamilie Helvetica ist eine serifenlose Linear-Antiqua mit klassizistischem Charakter und gehört zu den am weitesten verbreiteten Groteskschriften. Die ersten Schriftschnitte wurden ab 1956 von Max Miedinger und Eduard Hoffmann gestaltet (→ S. 51, 82).

Born in the German State of Saarland, *Andrea Tinnes*[1] (*1969) studied Communication Design at the University of Applied Science in Mainz and obtained her Master of Fine Arts for Graphic Design at the California Institute of the Arts. In 2004 she founded the font label »Typecuts«[2] in Berlin. From 2003 to 2008 Andrea Tinnes has been teaching Type Design and Typography at the Bergen National Academy of the Arts in Norway, and since 2007 at Burg Giebichenstein University of Art and Design in Halle / Germany. Her range of typefaces ranges from the ornamental Volvox[3] via a number of experimental fonts to type families like Skopex.[4] Her work was published, among others, in »Graphic Design for the 21st Century« and in »Eye Magazine« and received several awards. (TDC53; Red Dot).

How does the Typo Scene manifest itself in Berlin?

at: The type designer scene in Berlin is large and very lively, there is a regular exchange among the designers. There is still some truth in Erik Spiekermann's[5] statement that »the Netherlands have one type designer per square meter«, but meanwhile more and more type designers from The Hague[6] or Reading[7] settle in Berlin once they are qualified. This makes for a certain competition, of course, but I think it is great to live in a city where such an exchange is possible. Many of those designers are friends, they did their Master's together, met at conferences or formed private connections. I myself, for example, met some very interesting designers through Verena Gerlach.[8] The common interest brings you together, you meet for dinner or at various events and you exchange ideas.

In retrospect, how do you feel about the time you spent studying in Mainz?

This may be a little delicate, but I am rather critical about my time in Mainz, although I have to stress that in those days it was a very different university. The situation in those days can in no way be compared with that of today. I think that almost the entire staff has changed.

When I started studying in Mainz at the end of the Eighties, the computer had just started its triumph. The digital revolution and the post modern era in the field of graphic design were the talk of the day. Emigre, Eye Magazine, later Ray Gun, Carson[9] and Typography Now, in those days the Bible for all students, had a great influence. This, however, was not reflected in the teaching or in the discourse at the university, which led to a certain frustration among the students. I missed the theoretical, perhaps also the philosophical discourse in this very practice-oriented programme, I missed the inspiration from our teachers. That was the reason why I went to CalArts in the United States.

Having said this, however, I want to stress that I learned a lot from Hans Peter Willberg.[10] My emphasis in those days was on book design, and from him I learned the essentials of good typography. Of course he saw things from the viewpoint of an expert who came from classic typography / book design, and who didn't initially show too great an interest in the new and exciting design trends of other countries. But I must say in his favour that towards the end of his teaching career he became very open and incredibly interested in experimental work in the areas of graphic design, typography and typefaces. This, I think, is most impressive and I would like to emulate him by always questioning and being open-minded, particularly as a teacher, and never getting set in my ways.

Did you find this kind of openness and the discourse with contemporary design while in California? And could you put your German classic and accurate design approach to good use?

People often accuse the teachers and students in the »Graphic Design« programme of CalArts of going for a very »stylistic« design, only concentrating on form and visual language. But in fact the teaching in design theory and history there is at a very high level. There is a lively exchange, both in theory and in practice, both historic and contemporary, about all aspects of typography and graphic design, reflecting about the discipline. I gained a lot of experience and many new impressions, and it was here

1 typecuts.com

2 Typecuts is an independent type and graphics label, launched in 2004 by Andrea Tinnes and specializing in the creation and production of contemporary typefaces.

3 Volvox is a type system with five styles by Andrea Tinnes, which was inspired by the aesthetics of scientific drawings. Its superimpositions facilitate the design of complex patterns and structures.

4 Skopex is a sans-serif type family with vertical alignment, which was developed by Andrea Tinnes and published in 2005 by Primetype. A characteristic of the PTL Skopex Gothic are the sloped terminals of the ascenders and descenders.

5 The internationally renowned graphic and type designer Erik Spiekermann (*1947) launched the design studio MetaDesign in Berlin in 1979, later with branches in San Francisco and Zurich. He is co-founder of the international type foundry FontShop and is selling his typefaces together with Joan Spiekermann and Neville Brody under FontFont. Among his best-known and most influential types is the FF Meta.

6 The Koninklijke Academie van Beeldende Künsten, Den Haag (Royal Academy of Art, The Hague, Netherlands) was founded in 1682. The Master programme for Type Design and Media was instituted in 2002.

7 Reading University is in Reading, UK. It was founded in 1926 and is famous for its Master Programme for Type design. Many of the designers mentioned in this publication have studied or taught there (→ S. 62).

8 Verena Gerlach (→ S. 128)

9 The US typographer, type designer and surfer David Carson (*1957) became well-known for his innovative magazine design and the use of experimental typography. His style in typography and layout influenced the Grunge-aesthetics. Carson was Art Director of the magazine Ray Gun until he launched his own studio in New York in 1995 (→ S. 76).

10 Hans Peter Willberg (1930–2003) was a German typographer, illustrator and book designer as well as lecturer and author of textbooks. He was the Manager of the Stiftung Buchkunst in Frankfurt am Main and 1975–1996 Professor for typography and book design at the FH Mainz. Among his most important book titles are »Lesetypografie«, »Erste Hilfe in Typografie« and »Wegweiser Schrift«.

that I came to grips with design history. During my time in California I got to know many exciting personages and their work in art and design who I was not familiar with until then.

I am a very meticulous type of person, and that's how I first did my design work in Mainz, very pedantic and obsessed with detail. This meticulousness I took with me, but of course hoping that I would get to know other points of view, that I would improve and as a designer become more open-minded in order to develop alternative viewpoints and a different attitude. However, it was exactly this German accuracy which the Californian teachers were interested in. And the question of how this classic attitude, this solid craftsmanship as taught by an eminent authority like Willberg, could be combined with a very free and pictorial treatment of type and typography – all this became an important topic. My formal approach to design became more experimental and maybe also more organic at CalArts, and so I was able to develop my own style. On the one hand it is very playful and concerned with detail, on the other hand it is accurate to the point of perfectionism.

Your fonts are very experimental, but also very well developed. What do you personally expect from a font?

That depends on the context. There are many different ways of developing a type, the playful or the experimental dispute with the contents, the design of display alphabets which have to be attractive in a very visual way, or the development of a textface. For those three categories, i.e. experimental, striking or functional, there are of course totally different requirements.

Concerning typefaces for texts, for me it is important that they have a good rhythm, and – I suppose this must be the foremost challenge for every type designer – that there is something individual about them. They should be witty and inventive, without pushing this too much into the foreground, otherwise there is the danger that one tires of them. In order to be useful, a typeface should be constructed in such a way, that you have a large range of possibilities, like ligatures, alternative characters, different sets of numbers – in short, everything that's possible with OpenType. With my recent type families it has always been my objective to provide a well developed type family, so that typographers would be able to put them to good use.

Could you give us a brief summary of the history of Roletta?

This type has in fact a longer life history, because its genesis took a number of years. The idea originated during the discussion about a pitch for Radio Berlin-Brandenburg. We wanted to create a Logo or, respectively, a corporate font expressing the contrasts, the diversity as well as a fusion between the two states. The postmodern idea of combining different characteristics within the type was shown by the combination of round and angular corners.

A further requirement was a sans-serif type which on the one hand would save space when shown on the screen, on the other hand should have a somewhat livelier rhythm in its minuscles, making them easy to differentiate and very legible. With this in mind I started to develop the RBB type. Initially it was just a rudimentary version with four different weights. The pitch, unfortunately, was not decided in our favour, and so the type went on ice for a few years, until I started working on this design again.

The postmodern approach of this RBB type with its round and angular corners, however, was now too much of a reminder of the Nineties. So I had to decide whether they were all going to be round or angular. After first straightening out the complete character set, I met with Ole Schäfer[11] and both of us came to the conclusion that the other way – all of it rounded – was much more interesting.

In 2005 I then started research on how many rounded types there were on the market, e.g. Sauna[12] by Underware[13] or Hermes[14] by Optimo.[15] I also looked at the Fleischmann,[16] the Cooper Black[17] and the Helvetica Rounded[18] in order to see how I could position myself with another rounded type. And indeed, at that time there was no large type family combining sans and serif. This was the initial spark for the idea of developing a truly comprehensive rounded family.

But the development of Roletta was further delayed by other projects and changes in my own career. And finally, when it was finished in 2010, the boom in rounded type (which culminated between 2006 and 2010) was almost over. Nevertheless this typeface was an important step for my own development, because I learned many different principles about type design in the process. And it still isn't over, because now the history of the Roletta continues as »unrounded, straightened« Burg Grotesque.

What makes the Roletta an eye-catcher, and what are its special forms?

I think it is the contrast and the discrepancy between small and large point sizes. As a small type you don't see it as a rounded type, just as a restrained sans-serif typeface. But with increasing size you see more and more of the details, it gets softer, friendlier and more playful. A special feature is that the upper case letters are rather narrow in proportion and – in relationship with the minuscules – look rather constructed. Which means, if set only in upper case letters, the Roletta is a very modular, constructed serifeless type. In contrast, the style of the lower case letters is much more playful and lively. In a text they have completely different rhythms. Thus, the type leaves a wide scope for design purposes.

Andreas Tinnes was interviewed by Yvonne Kümmel and Jens Giesel.

11 Ole Schäfer (*1970) is a German typographer and type designer. He was involved in the development of the ITC type Officina, the FF Meta as well as the FF Info. In 2000 he published his FF Fago, two years later he established his type foundry and service company Primetype (→ S. 128).

12 The Sauna is a serifless type family designed by Bas Jacobs, Akiem Helmling and Sami Kortemäki. It was published in 2002 by Underware and consists of 18 styles and three weights.

13 Underware is a type foundry and a studio for graphic design, set up in 1999 by Akiem Helmling, Bas Jacobs and Sami Kortemäki. Not only does it produce fonts, but it also launched Typeradio, a radio programme on the subject on type.

14 Hermes is a new interpretation of the typeface used in the Hermes 3000 typewriter. Its simple geometric forms and rounded endings give the type a warm and friendly modernism. It was published in 2002 by the Graphic design studio Gavillet & Rust at Optimo.

15 Optimo is an independent type foundry in Geneva, Switzerland, founded in 1998 by Gilles Gavillet and David Rust. It develops and sells high quality type, e.g. type families like Theinhardt, Hermes, DidotElder, Dada Grotesk and Stanley.

16 The type family DTL Fleischmann goes back to the historic type styles by Johann Michael Fleischmann and incorporates text and display styles. The Leipzig designer Erhard Kaiser was strongly influenced by the original lead type face and published the Antiqua typefaces in 1997 at the Dutch Type Library.

17 In 1921 Oswald Bruce Cooper designed a fat type with rounded serifs for a poster. He published the type in 1922 by Barnhart Brothers & Spindler. The Cooper Black conformed to the Zeitgeist in advertising of those days: simple, friendly, bold.

18 The Helvetica type family is a serifless Linear-Antiqua with a classicistic character and is one of the most frequently used Grotesques. The first styles were designed from 1956 onwards by Max Miedinger and Eduard Hoffmann (→ S. 54, 87).

CALL FOR TYPE.

Neue Schriften.
New Typefaces.

190 **Actus**
Sonja Jenni

191 **Ambicase Fatface**
Craig Eliason

192 **Angata**
Julie Janet Chauffier

193 **Aniuk**
Thomas Gabriel

198 **BTP**
Guillaume Grall
Jeremy Perrodeau

199 **Canary**
Mark Frömberg

200 **Capricorne**
Christine Jungo

201 **Capucine**
Alice Savoie

206 **Crack**
Philipp Herrmann

207 **CA Cula**
Thomas Schostok

208 **Effra**
Jonas Schudel

209 **Faustina**
Alfonso Garcia

214 **Hashar**
Daniel Sabino

215 **Henriette**
Michael Hochleitner

216 **Herb**
Tim Ahrens

217 **Ingeborg**
Michael Hochleitner

222 **Marianne**
Benoît Bodhuin

223 **Mevum**
Angelo Stitz

224 **Modular**
Fabian Widmer

225 **Moskau Grotesk**
Björn Gogalla

230 **Premiéra**
Thomas Gabriel

231 **Quijote**
Erik Schöfer

232 **Retour**
Johannes Hucht

233 **Reuter Mono**
Peter Brugger

238 **Udine**
David Dusanek

239 **ZigZag**
Benoît Bodhuin

a

194 **Aria Pro**
Rui Abreu

a

195 **Azo Sans**
Rui Abreu

a

196 **Bernini Sans**
Tim Ahrens

a

197 **Blamage**
Phillipp Majdamin

A

202 **Carrosserie**
Fabian Widmer

a

203 **Clavo**
Michał Jarociński

a

204 **Colvert**
Chuvatin, Perez, Sarkis, Vlachou

a

205 **Cordale Arabic**
Ron Carpenter

A

210 **Fleurie**
FLAG

a

211 **Fonster**
Kathrin Esser

a

212 **Fry**
Oleg Macujev

a

213 **Glückskind**
Carmen Mauerer

a

218 **Instant**
Jérôme Knebusch

a

219 **Joos Pro**
Laurent Bourcellier

a

220 **Mainzer Grotesk**
Steffen Meyer

a

221 **Mantika Sans**
Jürgen Weltin

a

226 **Multiple Sans**
Marc Schütz

a

227 **Nord**
Julian Schambach

A

228 **Pavo**
Felix Beckheuer

a

229 **Planeta Pro**
Daniel Klauser

a

234 **Rooney Pro**
Jan Fromm

a

235 **Serendip Latin**
Rafael Saraiva

a

236 **Telegramo**
Laurenz Feinig

a

237 **Téras**
Sebastian Losch

abcdefghijklmnopqrstuvwxyz
ABCDEFGHIJKLMNOPQRSTUVWXYZ
0123456789

Actus

Gestalter. Designer. Sonja Jenni
Jahr. Year. 2012
Webseite. Website. xarten.ch

Mit ihrem markanten und zeitgenössischen kalligrafischen Stil greift die Actus zurück auf Vorbilder aus der Schriftgeschichte. Sie basiert auf charakteristischen Elementen der venezianischen Renaissance-Antiqua, die als Inbegriff guter Lesbarkeit gilt. Mit der Actus ist eine Schrift entstanden, die dennoch aktuell und modern erscheint.

Besonderheiten der Schrift: Die Versalhöhe ist deutlich niedriger als die Oberlänge, die relativ große x-Höhe verbessert die Lesbarkeit in kleinen Schriftgraden, die Serifen sind asymmetrisch, der Unterschied zwischen Haar- und Grundstrich ist relativ gering, die Innenformen der Schriftzeichen weisen Formen des kalligrafischen Schreibens auf und erinnern an das Schreiben mit schräg angesetzter Breitfeder, der Querstrich des e liegt leicht schräg und die Schriftachse ist bei Rundformen geneigt.

Die Schrift ist dank ihrer prägnanten, aber nicht aufdringlichen Einzelbuchstaben gut lesbar. (sj)

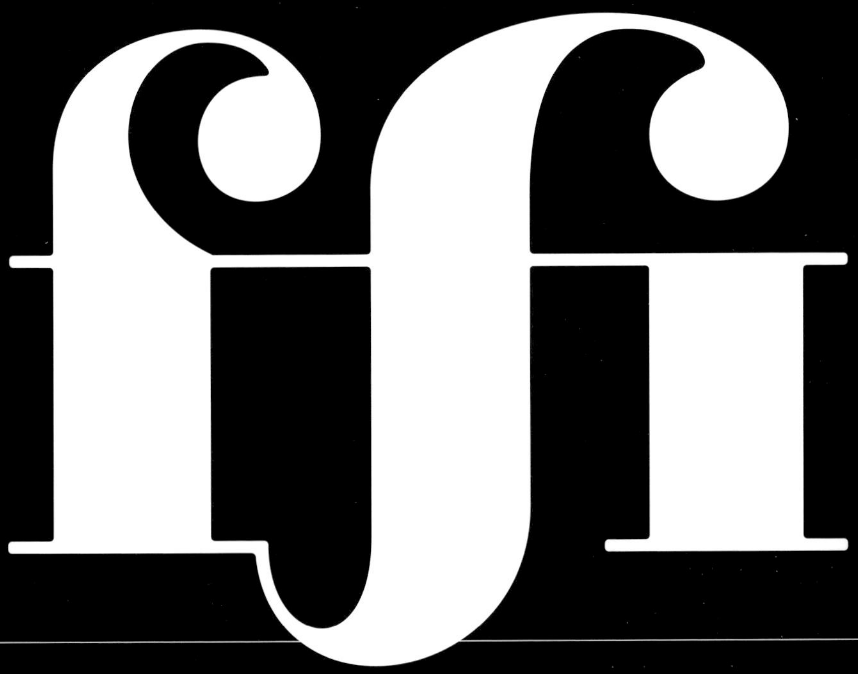

ABCDEFGHIJKLMN
OPQRSTUVWXYZ
0123456789

ABCDEFGHIJKLMN
OPQRSTUVWXYZ
0123456789

ABCDEFGHI
JKLMNOPQR
STUVWXYZ

ABCDEFGHI
JKLMNOPQR
STUVWXYZ

AMBICASE FATFACE

Gestalter. Designer. Craig Eliason
Label. Foundry. Teeline Fonts
Jahr. Year. 2011
Webseite. Website. myfonts.com

Ambicase Fatface takes an inventive approach to unicase font design, offering not »either/or«, but rather »both/and«. Each letter in Ambicase Fatface is a combination of its traditional upper- and lowercase forms, in an extra bold, high-contrast style.

Ambicase Fatface stands out as a carefully crafted experimental font: its eccentric forms do not hinder its readability. It is suitable for high-style display settings.

Ambicase Fatface offers a large character set and extensive OpenType features. Most notably, in modern OpenType-aware applications, Ambicase Fatface can be set in swash mode, which features sophisticated decorative flourishes that differ depending on whether the letter is at the beginning, middle, or end of a word.

Ambicase Fatface is available in two optical sizes: Regular and Poster. At very large sizes, the Poster cut, with its finer details, is recommended. (ce)

abcdefghijklmnopqrstuvwxyz
ABCDEFGHIJKLMNOPQRSTUVWXYZ
0123456789

Angata

Gestalter. Designer. Julie Janet Chauffier
Label. Foundry. University of Reading MATD
Jahr. Year. 2012
Webseite. Website. chauffier.com

Angata is an attempt to square the circle, or rather to semi-serif the sans. Its shapes are stretched over a conceptual frame: semi-serifs, large x-height, open squarish counters, proportional letters, sharp stroke modulation. To counter-balance the semi-serifs, its axis is somewhat off-kilter, keeping the reader alert. Thanks to its lack of true vertical and horizontal lines, Angata has a rough look. Its designed negative space mobilises individual letters into collective action; the typographic equivalent to setting your text to a marching tune.

Angata attempts to combine a contemporary look with familiar features, traditionally associated with comfortable reading. Angata means forward or »let's go« in Bambara, putting text in motion and galvanising readers. (jjc)

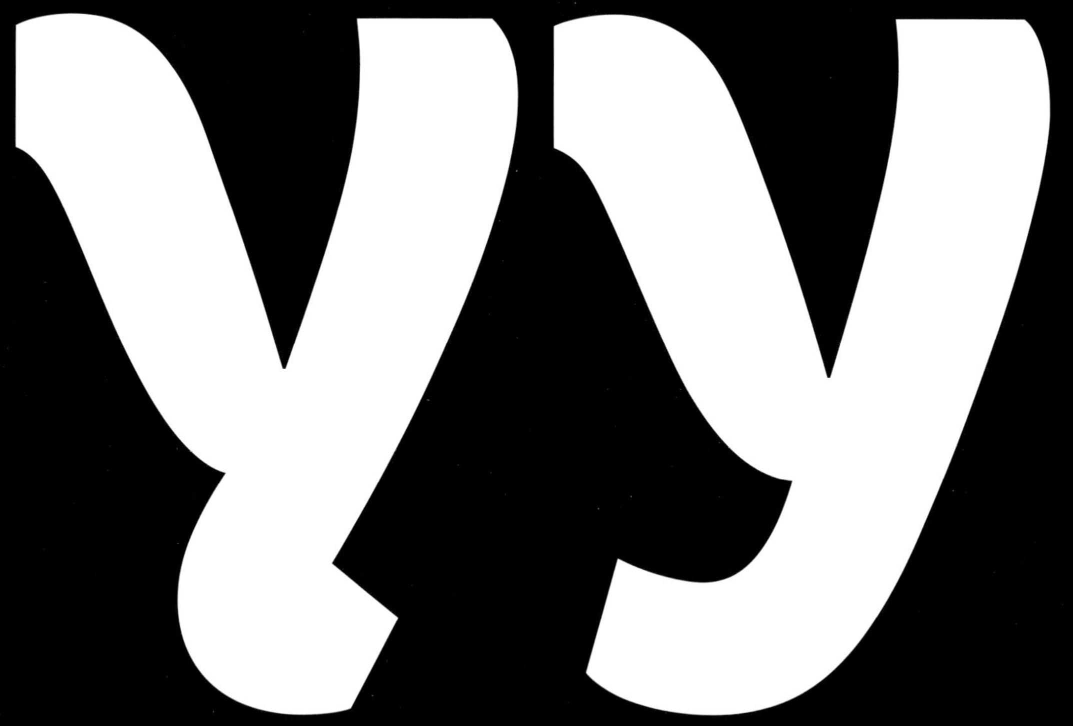

ANIUK

Aniuk Black
Aniuk Heavy

Aniuk Bold
Aniuk Medium

Aniuk Regular

abcdefghijklmnopqrstuvwxyz
ABCDEFGHIJKLMNOPQRSTUVWXYZ
0123456789

Aniuk

Gestalter. Designer. Thomas Gabriel
Label. Foundry. Typejockeys
Jahr. Year. 2010
Webseite. Website. typejockeys.com

Aniuk is a new original display type family designed and optimized for the use in large sizes. With five robust weights – Regular, Medium, Bold, Heavy and Black – it is perfectly suited for editorial, posters or logo design. A perfect make this a strong but playful typeface; a solid partner for your creative adventures.

Aniuk is lively, young, and probably a little crazy. However, there certainly is one thing that it is not: boring. (tg)

abcdefghijklmnopqrstuvwxyz
ABCDEFGHIJKLMNOPQRSTUVWXYZ
0123456789

Aria Pro

Gestalter. Designer. Rui Abreu
Label. Foundry. Fountain Type
Jahr. Year. 2011
Webseite. Website. r-typography.com

The inspiration for this typeface came from the epigraph on a frame of a nineteenth century painting. I was fascinated by the peculiar capitals of the inscription. The high contrast, and the overall quirkiness, especially the tail of the R and the oblique stems on the M, was interesting. I decided to draw a display font with high contrast and a vertical axis, in a reference to the transitional form. Still I wanted to capture the spirit of the original letters, which to me are so imbued with Romanticism. This approach allowed for some exuberance on the regular style, but also led to more calligraphic letterforms in the italic — in which »the flow of the curves« lead the way. To add to this epigraphic nature there is a number of ornaments that accompany words accordingly to their uppercase or lowercase form. For versatility there's also a good amount of ligatures, alternative glyphs, and a special set of ornamental numbers. (ra)

AZO SANS UBER
Azo Sans Black, *Italic*
Azo Sans Bold, *Italic*
Azo Sans Medium, *Italic*
Azo Sans Regular, *Italic*
Azo Sans Light, *Italic*
Azo Sans Thin, *Italic*

abcdefghijklmnopqrstuvwxyz
ABCDEFGHIJKLMNOPQRSTUVWXYZ
0123456789

Azo Sans

Gestalter. Designer. Rui Abreu
Label. Foundry. Rui Abreu
Jahr. Year. 2013
Webseite. Website. r-typography.com

Azo Sans is a grotesque loosely based on the elementary forms of geometry. It is inspired by the constructivist typefaces of the 1920s, but is instilled with a humanistic quality. With attributes usually found on dynamic sans-serifs, like the modulation of the strokes, humanistic proportions and open curved shapes, Azo Sans still feels geometrically constructed, somehow mechanical and precise, although geometry is hardly found. The o has the appearance of a perfect circle, however it is optically adjusted. The h, n, m have curves leaning to the right for a more dynamic look, rather than featuring symmetric counters with round junctures, like the typical circle based alphabet. Azo Sans is full of nuances that soften the strictness of pure geometry, making it more human and pleasant to read in longer body text, while maintaining a sober and rational appearance. Azo Sans comes in six weights from Thin to Black, with matching cursive italics. (ra) aagg

Bernina Sans Extrabold, *Italic*
Bernina Sans Bold, *Italic*
Bernina Sans Semibold, *Italic*
Bernina Sans Regular, *Italic*
Bernina Sans Light, *Italic*

Narrow Extrabold
Narrow Bold
Narrow Semibold
Narrow Regular
Narrow Light

Condensed Extrabold
Condensed Bold
Condensed Semibold
Condensed Regular
Condensed Light

Compressed Extrabold
Compressed Bold
Compressed Semibold
Compressed Regular
Compressed Light

abcdefghijklmnopqrstuvwxyz
ABCDEFGHIJKLMNOPQRSTUVWXYZ
0123456789

Bernini Sans

Gestalter. Designer. Tim Ahrens
Label. Foundry. Just Another Foundry
Jahr. Year. 2012
Webseite. Website. justanotherfoundry.com

JAF Bernini Sans was designed to speak with a clear and definite voice but without commenting on the text. In situations when the content is heterogeneous or unknown — such as in news websites or papers — the parametric extremes such as compressed or extrabold can give headlines a strong visual presence without stylistically interfering with the articles or images.

From a formal point of view, the family is inspired by the sturdy, unpretentious of Franklin Gothic and its »sculpted«, rather than pen-based or monolinear approach. It features open apertures in letter such as a, c, e, g, s, but combines them with round counters and shoulders flowing in. This formal paradox is rare but not unseen, most notably in Frutiger.

The family includes two fonts: JAF Bernino Sans and his sister JAF Bernina Sans — a more playful version with alternate shapes such as round dots and a double-storey g. The alternates are also available as OTF features so you can mix and match them without switching fonts. (ta)

abcdefghijklmnopqrstuvwxyz
ABCDEFGHIJKLMNOPQRSTUVWXYZ
0123456789

Blamage

Gestalter. Designer. Phillipp Majdamin
Jahr. Year. 2012
Webseite. Website. bblamage.net

Der Grundgedanke bei dieser Schrift war es, eine möglichst neutrale, humanistische Antiqua zu gestalten, die in größeren Punktgrößen deutlich ihren eigenen Charakter offenbart. Dafür wurden alle Serifen, Abstriche und Tropfen extrem spitz auslaufend gezeichnet. Es ging hauptsächlich darum, herauszufinden, wie weit man gehen kann und an welchen Stellen Kompromisse nötig sind. In erster Linie ist diese Schrift ein Experimentierfeld, aber durchaus mit dem Anspruch, auch allgemein im Mengentext zu funktionieren. (pm)

abcdefghijklmnopqrstuvwxyz
ABCDEFGHIJKLMNOPQRSTUVWXYZ
0123456789

BTP

Gestalter. Designer. Guillaume Grall
Jeremy Perrodeau
Label. Foundry. A is for Apple
Jahr. Year. 2012
Webseite. Website. aisforapple.fr

Building. In August 2011, the #195 issue of Étapes magazine, »Somewhere between graphic design and architecture ...«, has been the experimental field for the creation and use of the BTP typeface (by the Équipe Type, Art Direction by Guillaume Grall and Étienne Hervy). The design of BTP is freely inspired by the default typeface used in the computer-assisted design software AutoCAD, mainly used by architects for the conception of graphic documents (plans, elevations, etc.).

Typing. Once the issue has been published, Jeremy Perrodeau and Guillaume Grall have been working together to enhance the typeface. After several months of construction, BTP, a font with angular curves, gained an autonomous life and is now available in one multifunction weight on the independent foundry, A is for Apple, run by Émilie Rigaud. (gg+jp)

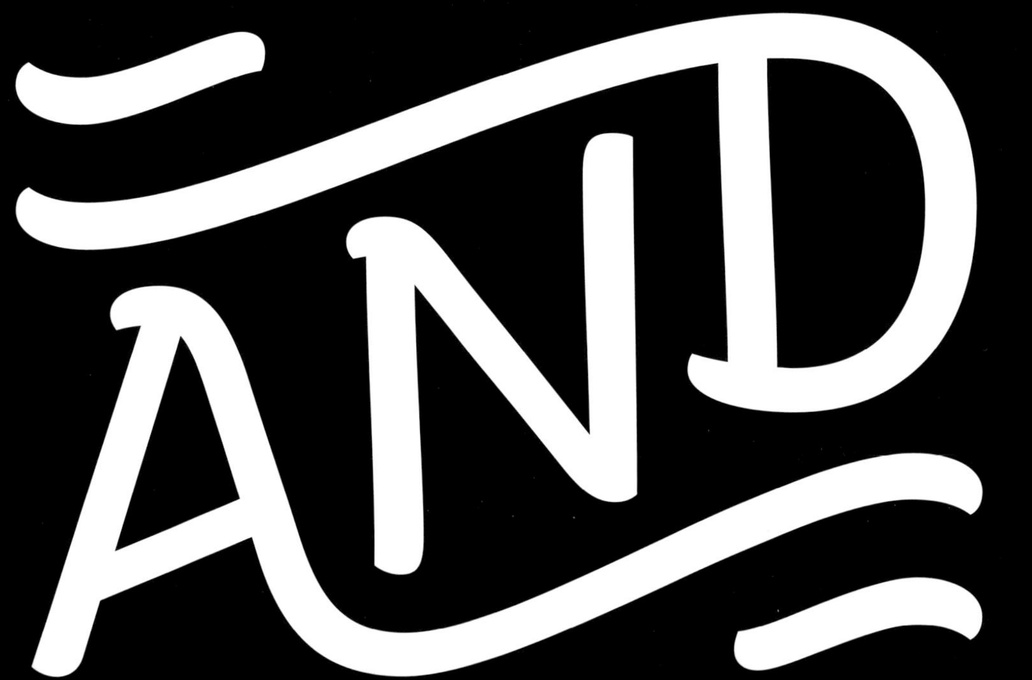

Canary Black
Canary Extra Bold

Canary Bold
Canary Medium

Canary Regular
Canary Light

abcdefghijklmnopqrstuvwxyz
ABCDEFGHIJKLMNOPQRSTUVWXYZ
0123456789

Canary

Gestalter. Designer. Mark Frömberg
Label. Foundry. Gestalten Fonts
Jahr. Year. 2012
Webseite. Website. mirque.de

Canary is a hybrid of left italic antiqua and brush painted script. It provides many automated and connectable letter substitutions that can slightly shift the font into the field of lettering rather than that of rigid and mundane typography.

Canary is designed to be compatible with a wide range of illustrative design practice. However, it works really well in medium-length copy settings as well. (mf)

CAPRICORNE

abcdefghijklmnopqrstuvwxyz
ABCDEFGHIJKLMNOPQRSTUVWXYZ
0123456789

abcdefghijklmnopqrstuvwxyz
ABCDEFGHIJKLMNOPQRSTUVWXYZ
0123456789

Capricorne

Gestalter. Designer. Christine Jungo
Jahr. Year. 2012 / 2013
Webseite. Website. crij.ch

Die Unterschiede der beiden Schnitte der Capricorne-Familie astro und terra werden erst bei größeren Schriftgraden sichtbar: Capricorne terra ist ein konstruierter Antiqua-Schnitt, der dank seiner horizontalen Ausrichtung, seiner kräftigen Schwärze und seinen geringen Ober- und Unterlängen auch in kleinen Schriftgraden und kompress gesetzt noch gut lesbar ist.

Das Verhältnis von rund zu eckig der Capricorne astro wird umgekehrt, indem die Bogen gebrochen werden. Die Proportionen und Strichkontraste werden optisch beibehalten. In kleinen Schriftgraden wirkt sie quasi identisch zur Antiqua, in großen Schriftgraden entfaltet sie ihre Kantigkeit. (cj)

CAPUCINE

Capucine Black, *Italic*
Capucine Bold, *Italic*
Capucine Regular, *Italic*
Capucine Light, *Italic*
Capucine Thin, Italic

abcdefghijklmnopqrstuvwxyz
ABCDEFGHIJKLMNOPQRSTUVWXYZ
0123456789

Capucine

Gestalter. Designer. Alice Savoie
Label. Foundry. Process Type Foundry
Jahr. Year. 2010
Webseite. Website. frenchtype.org

Although Capucine defies traditional categorization, it sits in a genre we are drawn to as users of type: a face with distinct personality able to straddle the worlds of both text and display with ease. In this context it should come as no surprise that its designer was born and raised in France, a country whose type history is rich with successful instances of such attempts. From Auriol and Grasset – typefaces that became symbolic of the Art Nouveau style – to the iconic designs of Roger Excoffon in the 1950s and 1960s, French type designers have often tried to fulfill the requirements of efficient text setting while retaining a gestural quality. Like many of its French predecessors, Capucine is driven by the eye rather than geometrical dogma, bringing a warmth and liveliness to the page. Designed to be both useful and friendly, Capucine carries a slight informal flavour due to its calligraphic roots. It is a lively design which makes for an expressive display face while remaining very legible at small sizes, thanks to its large x-height. With its 10 weights – all including small caps – the family offers a large typographic palette. From the delicacy of the thin to the generous curves of the bolder weights, Capucine is a versatile choice and contributes to produce memorable pieces of graphic communication. (as)

CARROSSERIE

CARROSSERIE FAT
CARROSSERIE BOLD
CARROSSERIE MEDIUM
CARROSSERIE REGULAR
CARROSSERIE LIGHT
CARROSSERIE EXTRALIGHT
CARROSSERIE THIN

ABCDEFGHIJKLMNOPQRSTUVWXYZ
0123456789

CARROSSERIE

Gestalter. Designer. Fabian Widmer
Label. Foundry. Letterwerk
Jahr. Year. 2010
Webseite. Website. letterwerk.ch

Carrosserie is made for display use, inspired by the shapes of the 1930s. It is a capital letter font with alternate characters and special domain symbols.

The font is available in Thin, Extra Light, Light, Regular, Medium, Bold and Fat. (fw)

Th

CLAVO

Clavo Black
Clavo ExtraBold
Clavo Bold
Clavo Medium

Clavo Regular
Clavo Book
Clavo Light
Clavo ExtraLight

Clavo UltraLight
Clavo Thin

abcdefghijklmnopqrstuvwxyz
ABCDEFGHIJKLMNOPQRSTUVWXYZ
0123456789

Clavo

Gestalter. Designer. Michał Jarociński
Label. Foundry. Dada Studio
Jahr. Year. 2013
Webseite. Website. dadastudio.pl

Clavo combines organic and industrial styles. Its warmth comes from subtle details, classical proportions and traditional forms, while harmonious structure prevents distraction in the reading process. This makes Clavo a universal typeface. In all sizes, from caption to display. The family consists of ten weights. They were not created in a linear way. The steps between the weights were adjusted carefully to avoid a mechanical graduation, in favour to optical harmony. Clavo covers all the 104 Latin languages. It contains a wide set of numerals, small capitals, fractions and other OpenType goodies. And of course every weight comes with matching italics. (mj)

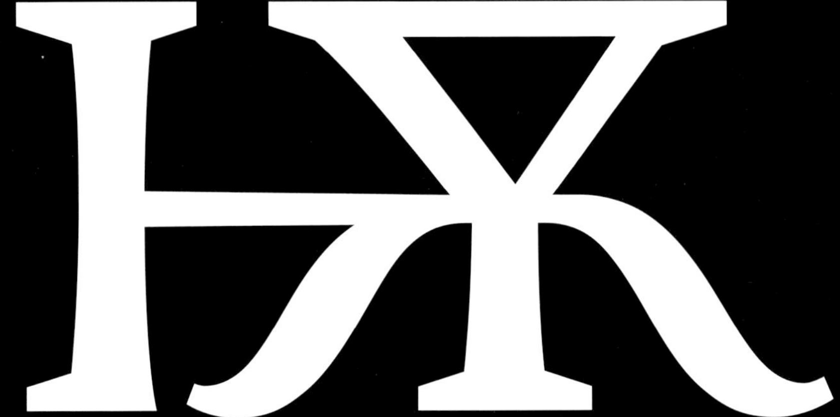

abcdefghijklmnopqrstuvwxyz
ABCDEFGHIJKLMNOPQRSTUVWXYZ
0123456789

αβγδεζηθικλμνξοπρστυφχψως
ΑΒΓΔΕΖΗΘΙΚΛΜΝΞΟΠΡΣΤΥΦΧΨΩ

абвгдеёжзийклмнопрстуфхцчшщъыьэюя
АБВГДЕЁЖЗИЙКЛМНОПРСТУФХЦЧШЩЪЫЬЭЮЯ

آ ب ت ث ج ح خ د ذ ر ز س ش
ص ض ط ظ ع غ ف ق ك ل م ن ه و ي ة

Colvert

Gestalter. Designer. Natalia Chuvatin, Jonathan Perez, Kristyan Sarkis & Irene Vlachou
Label. Foundry. typographies.fr
Jahr. Year. 2012
Webseite. Website. typographies.fr

Colvert includes four families: Colvert Arabic, Colvert Cyrillic, Colvert Greek and Colvert Latin. These four type families can be used alone, or blended with one another, in an harmonious way. Each family has been made by a designer and native speaker of the concerned writing system.

Each of the four families that make up Colvert is as visually differentiated as possible in order to express best the characteristics of the writing systems. (jp)

Cordale Arabic Bold, *Italic* Cordale Arabic Regular, *Italic*

abcdefghijklmnopqrstuvwxyz
ABCDEFGHIJKLMNOPQRSTUVWXYZ
0123456789

آ ب ت ث ج ح خ د ذ ر ز س ش
ص ض ط ظ ع غ ف ق ك ل م ن ه و ي ة

Cordale Arabic

Gestalter. Designer. Ron Carpenter
Label. Foundry. Dalton Maag
Jahr. Year. 2013
Webseite. Website. daltonmaag.com

Cordale Arabic is a natural addition to the Cordale family that works in harmony with the existing Latin script. The important features of Cordale's persona were incorporated into the Arabic design so that it retained its ability to be a strong workhorse. Cordale's distinctive serifs were continued through to the Arabic letterforms on the ascenders and the open character shapes were also retained. The Arabic is a Naskh style, which seemed to be the best fit for the hard working, but contemporary, details of Cordale's design. Every detail of the Arabic script was carefully considered to make it as legible as possible. (tt)

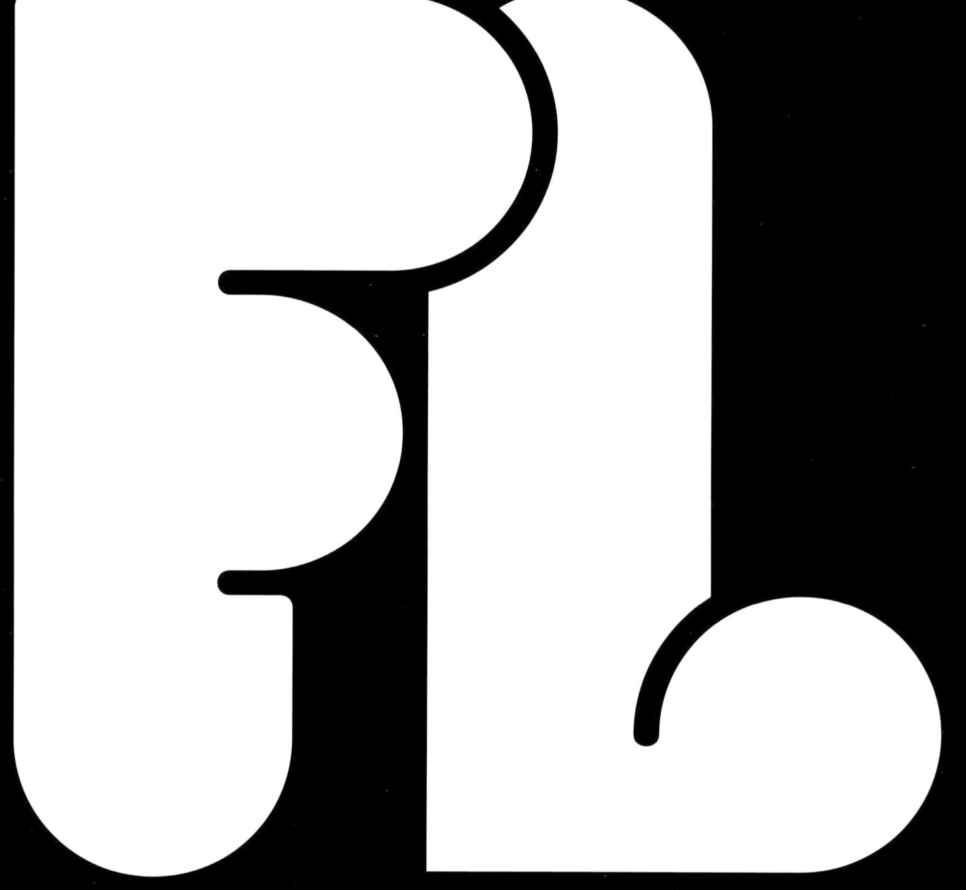

ABCDEFGHIJKLMNOPQRSTUVWXYZ
0123456789

CRACK

Gestalter. Designer. Philipp Herrmann
Label. Foundry. Fontseek.info
Jahr. Year. 2011
Webseite. Website. philippherrmann.ch

Erwin Poell, a designer from Heidelberg, Germany designed Poell Outline, Black and Shaded in 1972 based on a strict grid system. Crack is a modern interpretation in the year 2011. The overlapping of elements is applied in a more consistent manner and thereby taken to the extremes.

The design itself is modernized by making the letters more roundish. For a clearer display in smaller point sizes, the rounded counters are slightly expanded. Humour finds its way in type design. (ph)

CA Cula Superfat, *Italic*
CA Cula ExtraBold, *Italic*
CA Cula Bold, *Italic*
CA Cula Regular, *Italic*
CA Cula Light, *Italic*

abcdefghijklmnopqrstuvwxyz
ABCDEFGHIJKLMNOPQRSTUVWXYZ
0123456789

CA Cula

Gestalter. Designer.	Thomas Schostok
Label. Foundry.	Cape Arcona Type Foundry
Jahr. Year.	2011
Webseite. Website.	by.ths.nu

CA Cula is standing in the tradition of cool tempered sans-serif typefaces like DIN. But at a closer look it reveals a tendency towards rounder reading-friendly forms. The denaturalized ink traps give CA Cula a very special and individual look in display sizes, whereas in smaller sizes the positive aspects of huge ink traps show effect. The text looks clean and bright without black dots in the typographic image. This makes CA Cula suitable even for longer text, while the bold weight makes pretty cool headlines. (ts)

Effra Heavy, *Italic*
Effra Bold, *Italic*
Effra Medium, *Italic*
Effra Regular, *Italic*
Effra Light, *Italic*

abcdefghijklmnopqrstuvwxyz
ABCDEFGHIJKLMNOPQRSTUVWXYZ
0123456789

Effra

Gestalter. Designer. Jonas Schudel
Label. Foundry. Dalton Maag
Jahr. Year. 2008
Webseite. Website. grotesque.ch

Die Effra hat ihre Wurzeln in den frühesten kommerziell gehandelten Schriften überhaupt, der Caslon Junior, Two Lines Sans Serif, aus dem Jahr 1816. Sie wurde in einer zeitgenössischen Formensprache wieder zum Leben erweckt und zum Standard-Zeichensatz erweitert. Mit ihren fünf Gewichten ist sie gut ausgebaut und sorgt damit für eine flexible und gut strukturierte Gestaltung von Textbotschaften.

Die klare Linienführung und die offenen Proportionen sind die augenscheinlichsten Details der Effra; ihre Kreisformen erinnern an eine geometrische Basis und wirken zeitgemäß. Wo traditionelle Groteskformen erwartet werden, zeigt die Effra weiche und moderne, humanistische Details.

Die humanistischen Elemente des Designs der Effra unterstützen die Lesbarkeit und die subtile Stärkenabstufung erlaubt es, mit der Effra Texte hierarchisch gut zu strukturieren. (js)

abcdefghijklmnopqrstuvwxyz
ABCDEFGHIJKLMNOPQRSTUVWXYZ
0123456789

Faustina

Gestalter. Designer. Alfonso García
Label. Foundry. Tipos del Oeste
Jahr. Year. 2012/2013

Faustina is a font, designed for newspaper print with rugged, constructed characters for low quality print. Large x-height, semi-condensed proportions plus open counterpunches are characteristics that make Faustina a font that works

Best selected at 8 to 10 pt for news in Spanish. (Most Latin newspapers use North-American or European fonts.) Bold, Italic and Display Regular with small caps are under construction. (ag)

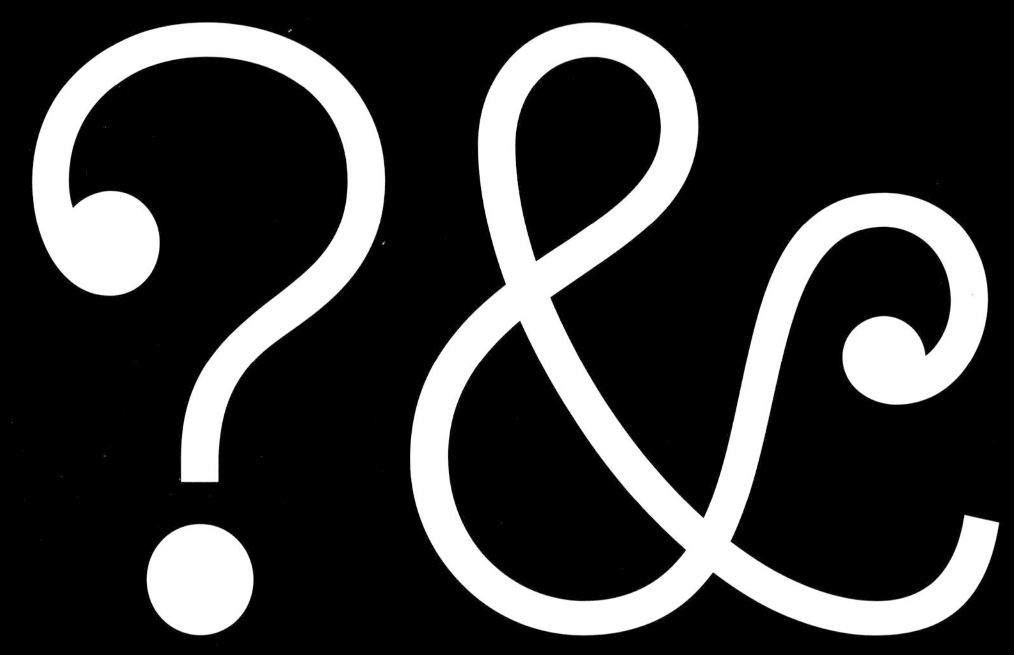

abcdefghijklmnopqrstuvwxyz
ABCDEFGHIJKLMNOPQRSTUVWXYZ
0123456789

abcdefghijklmnopqrstuvwxyz
ABCDEFGHIJKLMNOPQRSTUVWXYZ
0123456789

Fleurie

Gestalter. Designer. FLAG (Bastien Aubry, Dimitri Broquard)
Label. Foundry. Fontseek.info
Jahr. Year. 2010
Webseite. Website. fontseek.info

Fleurie Regular + Bold designed in collaboration with Aurèle Sack in 2009. FLAG is a graphic design studio established in 2002 by Bastien Aubry (*1974) and Dimitri Broquard (*1969) in Zurich / Switzerland. (ba+db)

abcdefghijklmnopqrstuvwxyz
ABCDEFGHIJKLMNOPQRSTUVWXYZ
0123456789

Fonster

Gestalter. Designer. Kathrin Esser
Label. Foundry. fonts.gestalten.com
Jahr. Year. 2011
Webseite. Website. kathrinesser.de

Fonster ist eine Antiqua in ihrer einprägsamsten Form. Ursprünglich als Headline-Schrift konzipiert, funktioniert sie aber auch problemlos als Fließtext-Schrift. Kennzeichnend für Fonster sind die Rundungen, welche jedoch tatsächlich eher eckig wirken – abgeleitet sind sie von gebrochenen Schriften.

Die Schrift Fonster entstand im 4. Semester an der FH Aachen im Cluster »Summer of Type« von Kai Oetzbach.

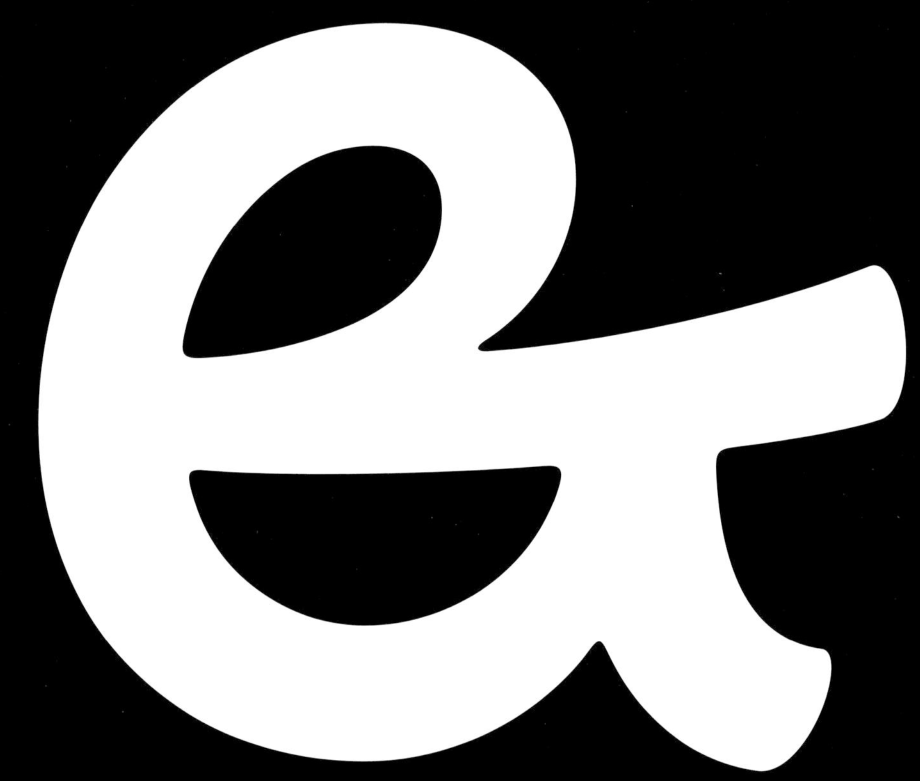

abcdefghijklmnopqrstuvwxyz
ABCDEFGHIJKLMNOPQRSTUVWXYZ
0123456789

Fry

Gestalter. Designer. Oleg Macujev
Label. Foundry. omtype
Jahr. Year. 2009
Webseite. Website. omtype.com

Fry was developed in 2008 as a corporate font for the Sky-Fish company (fish and seafood dealer). It is designed for small grades and has a friendly and fairytale historic flavour (can be seen in several archaic forms of Cyrillic de and ze). Fry takes openness and dynamism of humanistic sans-serif, simplicity and softness of lubok's letters (primitive style) and fluidity of shallow marine fry. Despite of funny style, Fry works well even in the 5 point size. In large sizes Fry demonstrates its originality, vivacity and softness, in small sizes the characteristics become less visible, and Fry's readability becomes more important. This makes the typeface suitable for many tasks of typography.

The font includes an extended set of Cyrillic and Latin, old style and lining figures, historical alternates. Fry was awarded for excellence in type design at Modern Cyrillic 2009 competition.
It received the second prize in display category at Granshan 2011. (om)

abcdefghijklmnopqrstuvwxyz
ABCDEFGHIJKLMNOPQRSTUVWXYZ
0123456789

Glückskind

Gestalter. Designer. Carmen Mauerer
Jahr. Year. 2012
Webseite. Website. carmenmauerer.de

Die Geschichte des »Glückskindes« begann in einem Schriftgestaltungskurs bei Dan Reynolds an der Hochschule Darmstadt, in dem Fotofunde von alten Schriften digitalisiert wurden. Sie basiert auf der Ladenbeschriftung einer alten Darmstädter Metzgerei. Nachdem alle vorhandenen Buchstaben nachgezeichnet, verändert, um weitere ergänzt und die Abstände zwischen ihnen optimiert waren, wurde die Schrift noch um Ligaturen und Alternativbuchstaben ergänzt, durch die sie einen möglichst gleichmäßigen und dennoch handgeschriebenen Charakter erhielt. Mit Ornamenten und Symbolen komplettiert, wurde aus dem »Glückskind« eine fröhliche Schrift, die sich gut auf Einladungen, Grußkarten, Verpackungen und als Logo- oder Titelschrift macht. (cm)

abcdefghijklmnopqrstuvwxyz
ABCDEFGHIJKLMNOPQRSTUVWXYZ
0123456789

Hashar

Gestalter. Designer. Daniel Sabino
Label. Foundry. Blackletra
Jahr. Year. 2012
Webseite. Website. blackletra.com

Hashar was designed to fill a gap among script typefaces: the absent of curves. It is the result of experiments with fractured curves and it offers some interesting solutions and a broad new feeling for the category.

Because script typefaces are generally very wavy and plenty of swashes and ornaments, they are normally associated with elegance and suavity. Hashar though is made to be agressive, suitable for other kind of projects. It is a true follower of Roger Excoffon's heritage but also has inspiration on the author's handwriting, on Blackletters and on graffiti works from São Paulo, Brazil. Hashar is a one weight font with many OpenType features, specially some swashes, stylistic and contextual alternates to play with. (ds)

Henriette, Black, *Italic*
Henriette Heavy, *Italic*
Henriette Bold, *Italic*
Henriette Medium, *Italic*
Henriette Regular, *Italic*

Condensed Black, *Italic*
Condensed Heavy, *Italic*
Condensed Bold, *Italic*
Condensed Medium, *Italic*
Condensed Regular, *Italic*

Compressed Black, *Italic*
Compressed Heavy, *Italic*
Compressed Bold, *Italic*
Compressed Medium, *Italic*
Compressed Regular, *Italic*

abcdefghijklmnopqrstuvwxyz
ABCDEFGHIJKLMNOPQRSTUVWXYZ
0123456789

Henriette

Gestalter. Designer. Michael Hochleitner
Label. Foundry. Typejockeys
Jahr. Year. 2012
Webseite. Website. typejockeys.com

In the 1920s the Viennese government decided to standardize the street signs across the city. A typeface was especially constructed for the purpose. It was available in a Heavy and a Bold Condensed version, to support short street names as well as longer ones. As the years went by, the typeface was adopted and redrawn by several enamel factories. These adaptations led to variations on the design, and to the fact that there isn't a Viennese street sign font but 16 – in part severely – different versions. Henriette is not a digitization of any of those versions; rather, it is influenced by all of them. The Italic versions are completely original and designed to accompany the Roman. (mh)

Herb Bold
Herb Regular

Herb Condensed Bold
Herb Condensed

abcdefghijklmnopqrstuvwxyz
ABCDEFGHIJKLMNOPQRSTUVWXYZ
0123456789

Herb

Gestalter. Designer. Tim Ahrens
Label. Foundry. Just Another Foundry
Jahr. Year. 2010
Webseite. Website. justanotherfoundry.com

JAF Herb is based on 16th century cursive broken scripts and printing types. Originally designed in the MA Typeface Design course at the University of Reading, it was further refined and extended in 2010.

The idea for JAF Herb was to develop a typeface that has the positive properties of black letter but does not evoke the same negative connotations – a type that has the complex, humane character of fraktur without looking conservative, aggressive or intolerant. (ta)

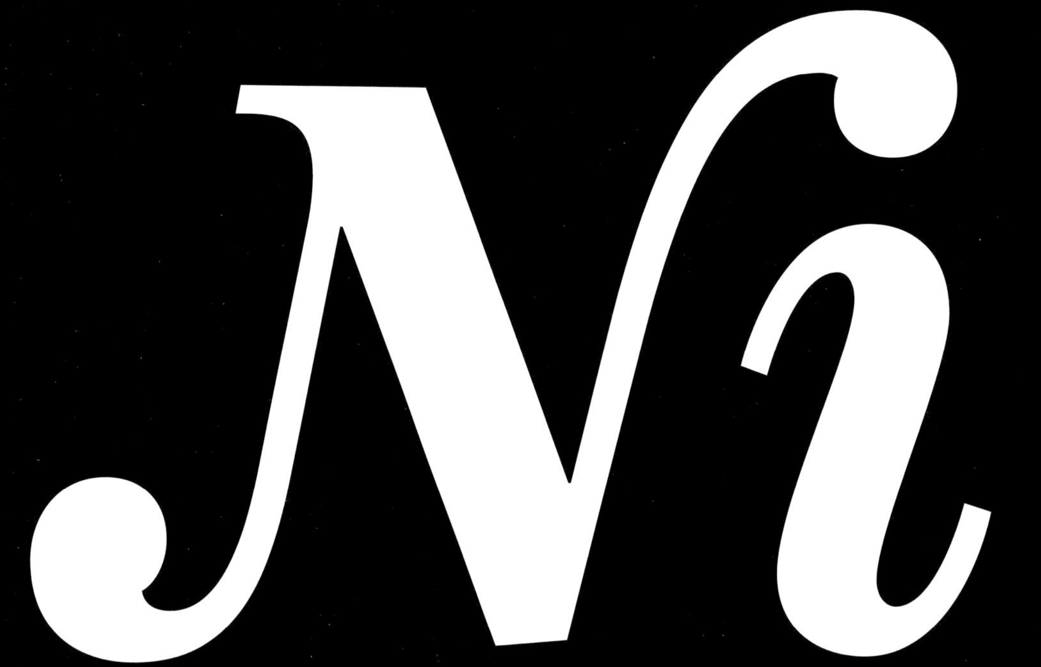

INGEBORG BLOCK
Ingeborg Fat, *Italic*
Ingeborg Heavy, *Italic*
Ingeborg Bold, *Italic*
Ingeborg Regular, *Italic*

abcdefghijklmnopqrstuvwxyz
ABCDEFGHIJKLMNOPQRSTUVWXYZ
0123456789

Ingeborg

Gestalter. Designer. Michael Hochleitner
Label. Foundry. Typejockeys
Jahr. Year. 2009
Webseite. Website. typejockeys.com

The Ingeborg family was designed with the intent of producing a readable modern face. Its roots might well be historic, but its approach is very contemporary. Ingeborg's Text Weights are functional and discreet. This was achieved without loosing the classic characteristics of a Didone typeface, which are the vertical stress and the high contrast. The display weights are designed to catch the reader's eye by an individual form language and a whole lot of ink on the paper. Nevertheless both are of one origin and work together in harmony. (mh)

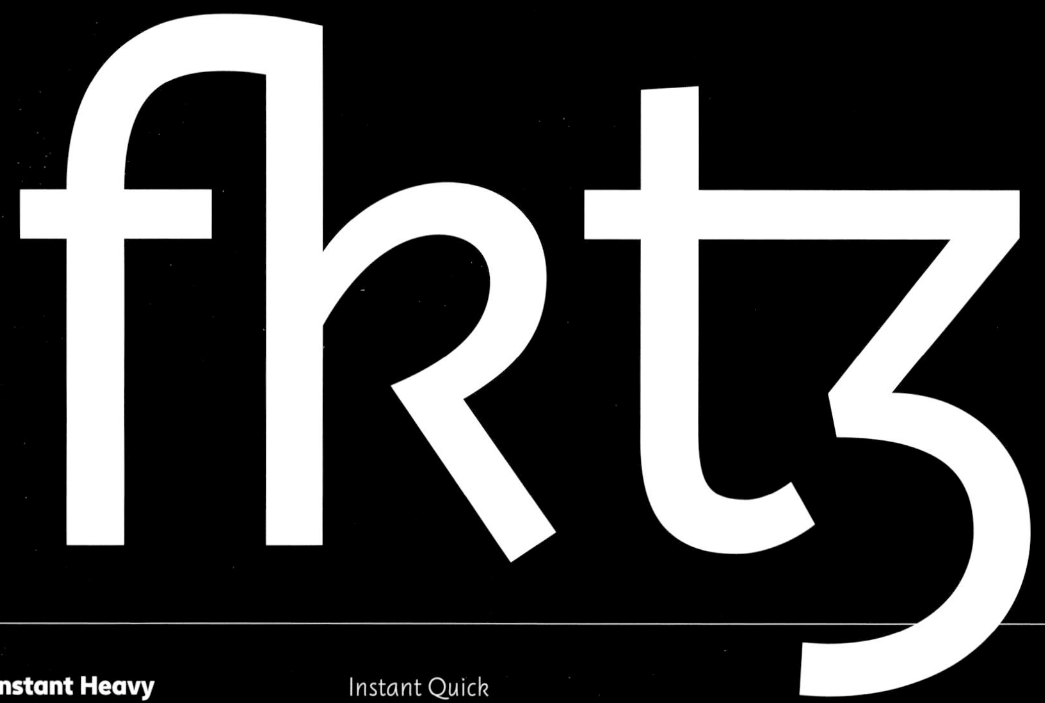

Instant Heavy
Instant Slow
Instant Regular
Instant Quick
Instant Vivid

abcdefghijklmnopqrstuvwxyz
ABCDEFGHIJKLMNOPQRSTUVWXYZ
0123456789

Instant

Gestalter. Designer. Jérôme Knebusch
Label. Foundry. Bureau des Affaires Typographiques
Jahr. Year. 2012
Webseite. Website. jeromeknebusch.net

Since its origins, typography is closely related to handwriting from which it originates. The quicker we write, the more the writing is done without raising the hand: typographic shapes derived from it are the so-called cursive shapes decomposed into several strokes match a slower execution speed? With speed comes thinness, and with slowness comes thickness. Thus was conceived Instant, a family in which each member is defined by a speed, which in return lends it both shape and weight: Instant Vivid, Instant Quick, Instant Regular, Instant Slow, Instant Heavy. Instant Viv is a fixation of handwriting, animated by a vivacious spirit, while Instant Heavy is a sans-serif, with robust and reassuring forms. Each weight is a intermediate step which Instant calls into question some established practices of typography. The linguistic part assigned to the Italic in relation with the Roman – usually highlighting a language element – is only possible by juxtaposing contrasted weights. On the other hand Instant allows to rediscover the first use of Italic, which was conceived as a design in itself, chosen for its own qualities. The wealth of expressiveness is at the heart of Instant which makes it possible for designers to stage a variety of connotations within a single text. (jk)

abcdefghijklmnopqrstuvwxyz
ABCDEFGHIJKLMNOPQRSTUVWXYZ
0123456789

Joos Pro

Gestalter. Designer. Laurent Bourcellier
Label. Foundry. typographies.fr
Jahr. Year. 2009
Webseite. Website. bourcellier.com

Joos Lambrecht, from Ghent, is one of the first important printers and punchcutters of the sixteenth century. He criticized frankly the reading habits and the typographical preferences of the Dutch and Flemish readers at that time. Since 1530 he tried to promote the use of roman types to replace black letter types, with little success. Lambrecht cut many roman types, but also a remarkable upright italic of which he was the only user. It is this Italic which inspired the Joos typeface.

This work is not a formal revival of Lambrecht's work, but faithfully fits into the scheme of its thought, which was to idealize roman types by bringing together the characteristic graceful shapes of italics and the angle of romans. Joos takes its inspiration in the principles of classical italics such as those of Francesco Griffo, but also from more contemporary shapes. In order to make the character optically vertical, it was necessary to work on each sign with a specific angle. Capitals all have a geometrical vertical stem, while the lowercases have an angle which vary between 0 and 2 degrees. (lb)

MAINZER GROTESK

Mainzer Grotesk Fett
Mainzer Grotesk Halbfett

Mainzer Grotesk Normal

abcdefghijklmnopqrstuvwxyz
ABCDEFGHIJKLMNOPQRSTUVWXYZ
0123456789

Mainzer Grotesk

Gestalter. Designer. Steffen Meyer
Label. Foundry. Zweizehn
Jahr. Year. 2013
Webseite. Website. zweizehn.com

The Mainzer Grotesk was designed as part of the BA-Thesis »Das Mainzer Grotesk Experiment« at the FH Mainz (coached by Prof. Dr. Isabel Naegele).

It brought in question the use of letterpress printing in today's digital era. The emphasis was rather on the artistic and experimental use of the letterpress instead of focusing on its plain retro charme. To testify whether or not these expensive and time-consuming form of hand printing has some value in our age, we created and cast our own letterpress type – the Mainzer Grotesk and brought it into use. The project was printed and published a year ago, in Gutenberg's hometown Mainz about 570 years after his great invention. The typeface itself derives from classical grotesque typefaces from the 1930s but has been improved for modern production techniques, milling the matrix with a CNC machine. This is were the Mainzer Grotesk's slightly rounded edges and its thickness derive from. (sm)

Mantika Sans Bold, *Italic*   Mantika Sans Regular, *Italic*

abcdefghijklmnopqrstuvwxyz
ABCDEFGHIJKLMNOPQRSTUVWXYZ
0123456789

Mantika Sans

Gestalter. Designer. Jürgen Weltin
Label. Foundry. Monotype
Jahr. Year. 2011
Webseite. Website. typematters.de

The aim of producing Mantika Sans was to create a typeface with excellent legibility even in small sizes not just by means of the x-height, which is tall in comparison with the capital letters, but also by using clearly defined and well differentiated designs for critical letters, such as i, I and l. The elaborately designed and highly individual set of italics are characterised by bevelled line endings and the slight variation in thickness of verticals, resulting in a very dynamic character. Short ascenders and descenders give it a compact appearance that is also underscored by its condensed proportions. The uppercase numerals are slightly shorter than the uppercase letters, ensuring that the latter can be sympathetically incorporated within continuous text. The Italics appear to be almost upright. Within the variety of forms of the Italics there are many contrasting terminal elements that create dynamism. The result is a diversity of interaction between the rounded and angular forms. (jw)

MARIANNE

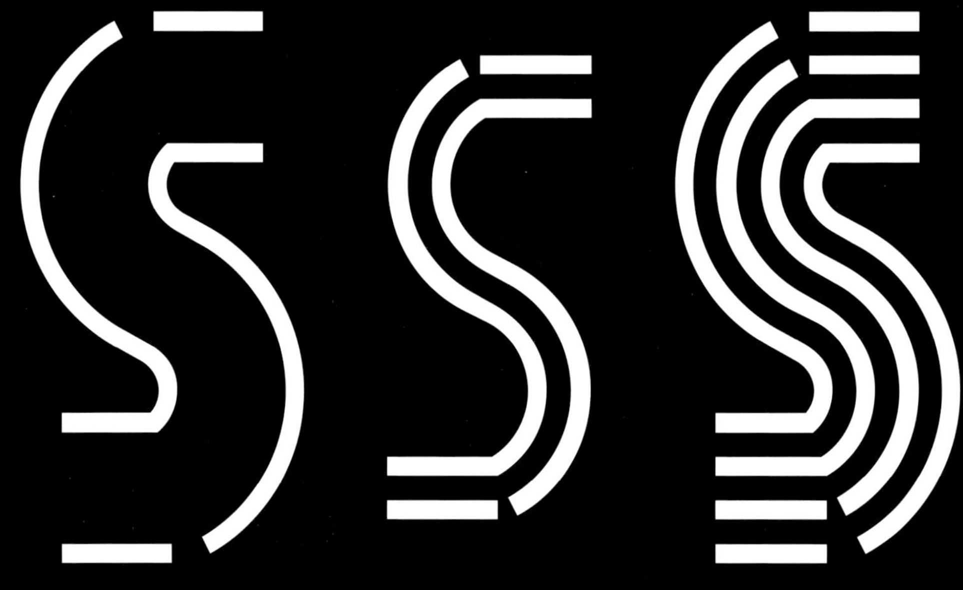

ABCDEFGHIJKLMNOPQRSTUVWXYZ
0123456789

ABCDEFGHIJKLMNOPQRSTUVWXYZ
0123456789

ABCDEFGHIJKLMNOPQRSTUVWXYZ
0123456789

MARIANNE

Gestalter. Designer. Benoît Bodhuin
Label. Foundry. benbenworld.com/fontes.htm
Jahr. Year. 2012
Webseite. Website. benbenworld.com

Marianne is a headline lineal and protest writing (caps only), made of tape modules joined by drawing a typical notch. Three styles – Inline, Outline and Solid – each with variants OpenType, many original ligatures (including HTTP ...) and alternative A leaning on his right leg, allow many combinations and uses. (bb)

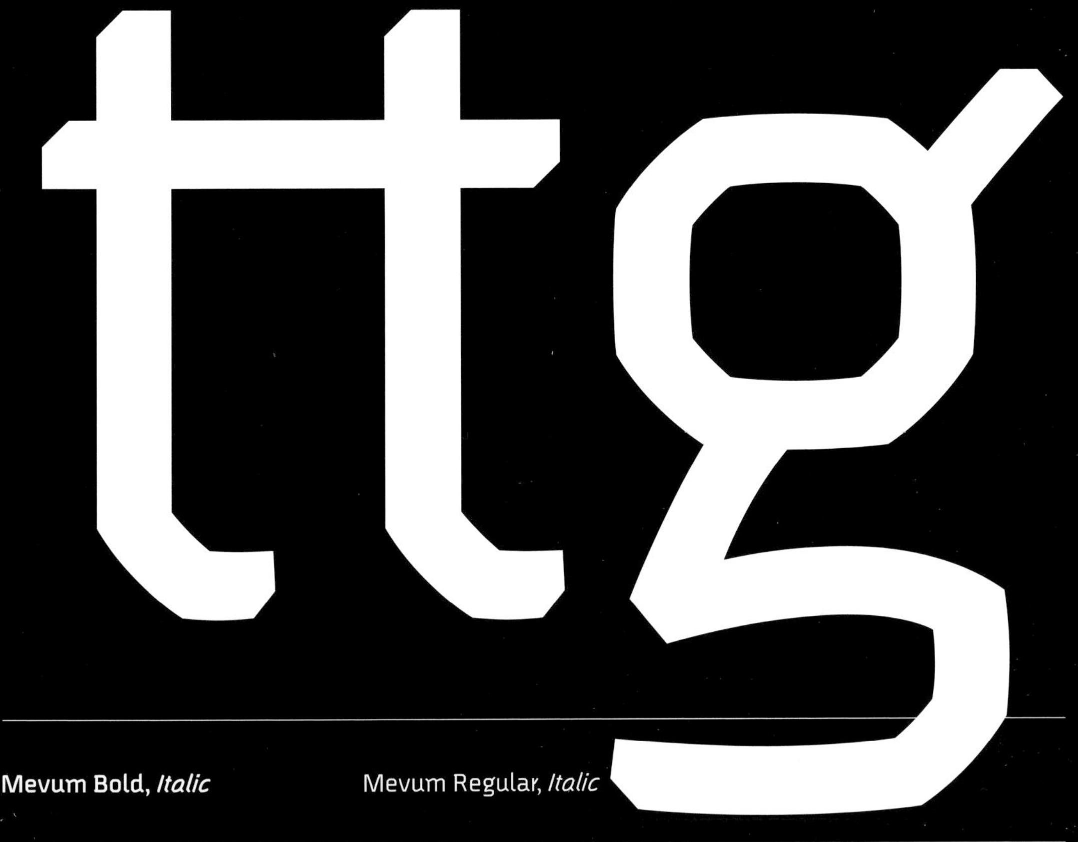

Mevum Bold, ***Italic***   Mevum Regular, *Italic*

abcdefghijklmnopqrstuvwxyz
ABCDEFGHIJKLMNOPQRSTUVWXYZ
0123456789

Mevum

Gestalter. Designer. Angelo Stitz
Label. Foundry. Gestalten Fonts
Jahr. Year. 2012
Webseite. Website. fonts.gestalten.com/mevum.html

Mevum belongs to an entire font family that corresponds to the ample Western-Latin letter set. Treating each letter as an individual element of a corporate logotype, the font's unique verve aims to create a simple, strongly grotesque but warm character. With the elegant combination of tough and minimalistic characteristics that are declined by its feminine undertone, Mevum makes a unique and fresh impression. The additional counterpart to the regular set consists of an extraordinary Italic. Both styles are available in bold weights and incorporate old style figures on its main glyph position, as well as lining figures, that are applicable to proportional and fixed widths. Its aesthetic structure can be valuable for setting either text or tables. Furthermore, the letter set contains eligible details like different ligatures such as the ß expedient for the German alphabet. (as)

Modular Slab Bold Modular Slab Roman Modular Sans Roman

abcdefghijklmnopqrstuvwxyz
ABCDEFGHIJKLMNOPQRSTUVWXYZ
0123456789

Gestalter. Designer. Fabian Widmer
Label. Foundry. Letterwerk
Jahr. Year. 2012
Webseite. Website. letterwerk.ch

Modular is a six layer stacking display typeface. Six different shapes fill counterpunches and spaces between the letters. Six different layers build the type. In Modular each layer can be colored differently. Furthermore it is possible to adjust the order of the layers as well as the »blend mode«. This gives you the opportunity to design a huge range of different moods. Scripts and actions for Adobe InDesign and Adobe Illustrator were programmed. It's pretty easy to create stacked Modular text in those programs – just load the actions and press play! Existing cuts: Modular Sans Roman (The shapes are exactly arranged), Modular Slab Roman (The Shapes do overlap and are out of alignment), Modular Slab Bold (The Shapes do overlap and are out of alignment). (fw)

Moskau Grotesk ExtraBold, *Italic*
Moskau Grotesk Bold, *Italic*
Moskau Grotesk Medium, *Italic*
Moskau Grotesk Regular, *Italic*

Moskau Grotesk Light, *Italic*
Moskau Grotesk Thin, *Italic*
Moskau Grotesk ExtraLight, *Italic*

abcdefghijklmnopqrstuvwxyz
ABCDEFGHIJKLMNOPQRSTUVWXYZ
0123456789

Moskau Grotesk

Gestalter. Designer. Björn Gogalla
Label. Foundry. Letter Edit
Jahr. Year. 2012
Webseite. Website. letteredit.de

Das Café Moskau, gegenüber dem Kino International in der Karl-Marx-Allee in Berlin-Mitte war eine der Prestigebauten der ehemaligen DDR. Der Namenszug auf dem Dach wurde von dem Grafiker Klaus Wittkugel entworfen. Die Beschilderung des Café Moskau, mit den in Versalien gesetzten Worten »RESTAURANT«, »CAFE«, »KONZERT«, »MOSKAU« und »MOCKBA«, bildet die Vorlage der Moskau Grotesk. Dabei versteht sich die Schrift nicht als Replik. Einige Unzulänglichkeiten wurden »ausgebessert«. Um die Gesamtcharakteristik bewahren zu können, wurde jedoch nicht völlig auf Eigenheiten verzichtet. Die Gemeinen und alle fehlenden Versalien sind komplett neu entworfen. Es ist nicht verwunderlich, dass die schlichte, unaufdringliche, geometrische Grundausrichtung der Schrift eine Brücke zur Architektur der 1960er Jahre schlägt. Inspiriert von den in dieser Zeit beliebten verschiedenen Möglichkeiten des architektonischen Musters und Wandreliefs entstanden ergänzend zwei Pattern-Fonts. (bg)

aaaa
aaaa

MULTIPLE SANS

abcdefghijklmnopqrstuvwxyz
ABCDEFGHIJKLMNOPQRSTUVWXYZ
0123456789

Multiple Sans

Gestalter. Designer. Marc Schütz
Jahr. Year. 2011
Webseite. Website. multiplesans.de

Multiple Sans is a conceptual sans serif typeface based on the idea that a font's glyph forms are balanced between unity and variety.

Each glyph possesses formal properties that have either a geometric or a humanist tendency. Properties that reduce formal variety and therefore support the even look of a typeface are of geometric nature, whereas the distinct elements of different characters are emphasized through humanist properties that make the font more legible.

The balance of these properties determines a specific position on the style axis whose extreme points are accordingly named after the opposing typeface groups geometric and humanist sans.

All glyphs of a font share common properties, which are called first grade properties. The first part of every type design process is the selection of these basic properties. Second grade properties only apply to various groups of letters and can be combined in different ways.

The integrative design process includes as many alternate forms as possible in one font. (ms)

abcdefghijklmnopqrstuvwxyz

ABCDEFGHIJKLMNOPQRSTUVWXYZ

0123456789

Nord

Gestalter. Designer. Julian Schambach
Jahr. Year. 2012
Webseite. Website. julianschamba.ch

Die Nord Italic wurde ursprünglich als Display-Font für ein Design-Projekt gestaltet und bestand nur aus Versalien. Erst in einem zweiten Schritt wurden die Gemeinen ergänzt. Die Nord Italic ist eine feingliedrige Schrift, die sowohl Schlichtheit als auch Sportlichkeit ausdrückt. Streng genommen ist der Zusatz »italic« nicht ganz richtig, da die kursiv gestellte Variante die Basis bildete und eine Aufrechte erst später als Familienmitglied hinzukam. (js)

La!

PAVO

Gestalter. Designer. Felix Beckheuer
Jahr. Year. 2013
Webseite. Website. felix.beckheuer.de

Die Pavo wurde inspiriert von Herb Lubalins Avant Garde. Ursprünglich 1968 für das Logo des gleichnamigen Kulturmagazins gezeichnet, wurde die Avant Garde Anfang der 1970er Jahre als Schrift ausgebaut.

Die Pavo zeichnet sich als Display-Schrift durch starke Strichstärkenkontraste aus und bietet mit 450 Glyphen, inklusive 170 Ligaturen, eine Vielzahl an Kombinationsmöglichkeiten.

Planeta Pro Plakat
Planeta Pro Bold

Planeta Pro Regular
Planeta Pro Light

abcdefghijklmnopqrstuvwxyz
ABCDEFGHIJKLMNOPQRSTUVWXYZ
0123456789

Planeta Pro

Gestalter. Designer. Dani Klauser
Label. Foundry. Dani Klauser Grafik Design
Jahr. Year. 2009 / 2013
Webseite. Website. dkgd.ch

The intention of creating Planeta is the transportation of beauty and the geometric simplicity of constructed typefaces. Planeta reinterprets the form of the 1920s and conveys its spirit with contemporary freshness. Planeta comes in four cuts. Light, Regular and Bold form the basis for the daily use. The Plakat-cut was drawn for character sizes over 100 points and is intended for the use in Poster design and big scaled letters.

In 2009 Planeta came out as a regular four-cut font. In spring 2013 a more usefull »Planeta Pro« version followed. New OpenType features, alternative characters and a full function figure-set come along with the »Planeta Pro«. (dk)

PREMIÉRA

kk!

Premiéra Bold Premiéra Book, *Italic*

abcdefghijklmnopqrstuvwxyz
ABCDEFGHIJKLMNOPQRSTUVWXYZ
0123456789

Premiéra

Gestalter. Designer.	Thomas Gabriel
Label. Foundry.	Typejockeys
Jahr. Year.	2009
Webseite. Website.	typejockeys.com

Premiéra is a book typeface specifically designed to work in small sizes. It is available in 3 weights: The Book for main text demands and Bold and Italic to create different kind of emphasis. A strong x-height and short ascender / descender make this a very legible and elegant typeface, very suitable for use in books and newspapers. The idea for Premiéra comes from a demand on developing a typeface that works very well in small prints. Its main features, straight lines and sharp forms developed through a process of testing readability in very small print sizes. The result is a typeface with a strong personality whether you read it in small or in bigger size. (tg)

abcdefghijklmnopqrstuvwxyz
ABCDEFGHIJKLMNOPQRSTUVWXYZ
0123456789

Quijote

Gestalter. Designer. Erik Schöfer
Jahr. Year. 2013

»Quijotes« Ansatz entstand aus der Beschäftigung mit der Herleitung von Buchstabenformen

Zunächst überraschender-, später logischerweise kamen in den Schreibübungen Formen zum Vorschein, die bereits über eine lange Geschichte verfügen, im Zuge der modernistischen Vereinfachung in Groteskschriften aber immer weniger

die zum einen historische Formen in eine geometrisch-konstruierte Grundarchitektur überführt, zum anderen Buchstabenformen schafft, »die wir« — wie ein lesen-lernender Erstklässler versicherte — »noch nicht hatten«. (es)

abcdefghijklmnopqrstuvwxyz
ABCDEFGHIJKLMNOPQRSTUVWXYZ
0123456789

Retour

Gestalter. Designer. Johannes Hucht
Label. Foundry. Avoid Red Arrows
Jahr. Year. 2012
Webseite. Website. johanneshucht.de

Retour ist eine ungewöhnliche Sans Serif-Schrift. Sie kombiniert den Charakter klassischer Kalligrafie-Schriften mit den stark variierenden Breitenunterschieden von konstruierten Groteskschriften. Charakteristisch für die Retour sind ihre schrägen Auf- und Abstriche und ihre extremen Strichstärkenunterschiede, die den Lettern eine leicht dreidimensionale Anmutung verleihen. (jh)

it

abcdefghijklmnopqrstuvwxyz
ABCDEFGHIJKLMNOPQRSTUVWXYZ
0123456789

Reuter Mono

Gestalter. Designer. Peter Brugger
Jahr. Year. 2010
Webseite. Website. peterbrugger.net

The typeface Reuter Mono was developed for the book »Hans Joachim Reuter – Leuchtende Bilder/Luminescent Images«. Reuter was an internationally renowned physician and scientist. He was a leading figure in the development of endoscopic imaging techniques in the 1960s and with his »nuclear medical image compositions« became a pioneer of Science Art in the 1970s. (pb)

Rooney Pro Black, *Italic*
Rooney Pro Heavy, *Italic*

Rooney Pro Bold, *Italic*
Rooney Pro Medium, *Italic*

Rooney Pro Regular, *Italic*
Rooney Pro Light, *Italic*

abcdefghijklmnopqrstuvwxyz
ABCDEFGHIJKLMNOPQRSTUVWXYZ
0123456789

Rooney Pro

Gestalter. Designer. Jan Fromm
Label. Foundry. Jan Fromm
Jahr. Year. 2010
Webseite. Website. janfromm.de

Das Hauptmerkmal der Rooney Pro sind die runden Abschlüsse der Serifen und Strichenden, die den Zeichen ein warmes und sympathisches Antlitz verleihen und die sich besonders in Display-Größen offenbaren. Durch die Formensprache klassischer Antiquas, die offenen Buchstabenformen, eine schräge Kontrastachse sowie einen moderaten Kontrast erhält die Rooney einen seriösen Charakter; eine große x-Höhe unterstützt zudem die Lesbarkeit.

Die Anwendungsmöglichkeiten der Rooney reichen vom Editorial Design über die Buchgestaltung bis hin zur Entwicklung von Logos und Produktverpackungen. Rooney besteht aus sechs Gewichten – Light, Regular, Medium, Bold, Heavy und Black – für jedes Gewicht gibt es eine passende Kursive. (jf)

Serendip Latin Heavy
Serendip Latin Regular, *Italic*

Serendip Latin Thin

abcdefghijklmnopqrstuvwxyz
ABCDEFGHIJKLMNOPQRSTUVWXYZ
0123456789

Serendip Latin

Gestalter. Designer. Rafael Saraiva
Jahr. Year. 2012
Webseite. Website. rafaelsaraiva.com

Serendip was discovered in a nine-month journey throughout seas never before navigated. The type family is intended to compose the canonical texts of Theravāda Buddhism, supporting Pāli language transliterated in Latin and Sinhala scripts.

Theravāda scriptures, known as Pāli Tipitaka, compile the teachings of the historical Buddha in extensive 45 volumes. The project brief was outlined to solve this editorial problem as a robust book typeface, highly legible for long run text setting. However, Serendip has extrapolated its original proposition with the addition of extreme weights and Serendip family is now a versatile system which is suitable for a wide range of editorial projects. (rs)

TELEGRAMO

telegramo Bold
telegramo Medium
telegramo Regular

abcdefghijklmnopqrstuvwxyz
ABCDEFGHIJKLMNOPQRSTUVWXYZ
0123456789

Telegramo

Gestalter. Designer. Laurenz Feinig
Label. Foundry. Volcano Type
Jahr. Year. 2010
Webseite. Website. volcano-type.de

Telegramo is modelled on a historic telegraph from Belgrade to Vienna 1914. The original archetypical character set consists of lowercase letters and numerals only. Uppercase letters and special characters were added after careful research. Contact pressure variations of the rudimentary type writing machine are directly imitated in the three weights: the regular weights edges are sharp, medium edges are rounded and the bold letters can nearly be called soft. Since the original typeface did not seem perfectly suitable for modern desktop publishing purposes, two additional stylistic sets were created for each weight, improving certain issues in rhythm, legibility and quirkiness. (lf)

abcdefghijklmnopqrstuvwxyz
ABCDEFGHIJKLMNOPQRSTUVWXYZ
0123456789

Téras

Gestalter. Designer. Sebastian Losch
Jahr. Year. 2013
Webseite. Website. sebastianlosch.de

Téras is a multi-script typeface for magazines, catalogues and other applications of complex typography. It unites Latin, Greek, Arabic and Tamil in stylistic harmony whilst respecting the cultural heritage of each script. The lack of vertical serifs emphasises the reading direction, enlarges the counters and thus invites the reader with open arms. Sharp corners and squarish forms give the text a crisp feeling which makes the typeface look like a bike after an accident – demolished yet still functioning and aesthetically pleasing. A large x-height and the generous use of white space make it legible in small sizes; unobstrusive details convey the highly individual character in larger sizes. Téras comes in four weights plus italics and supports a variety of OpenType features. (sl)

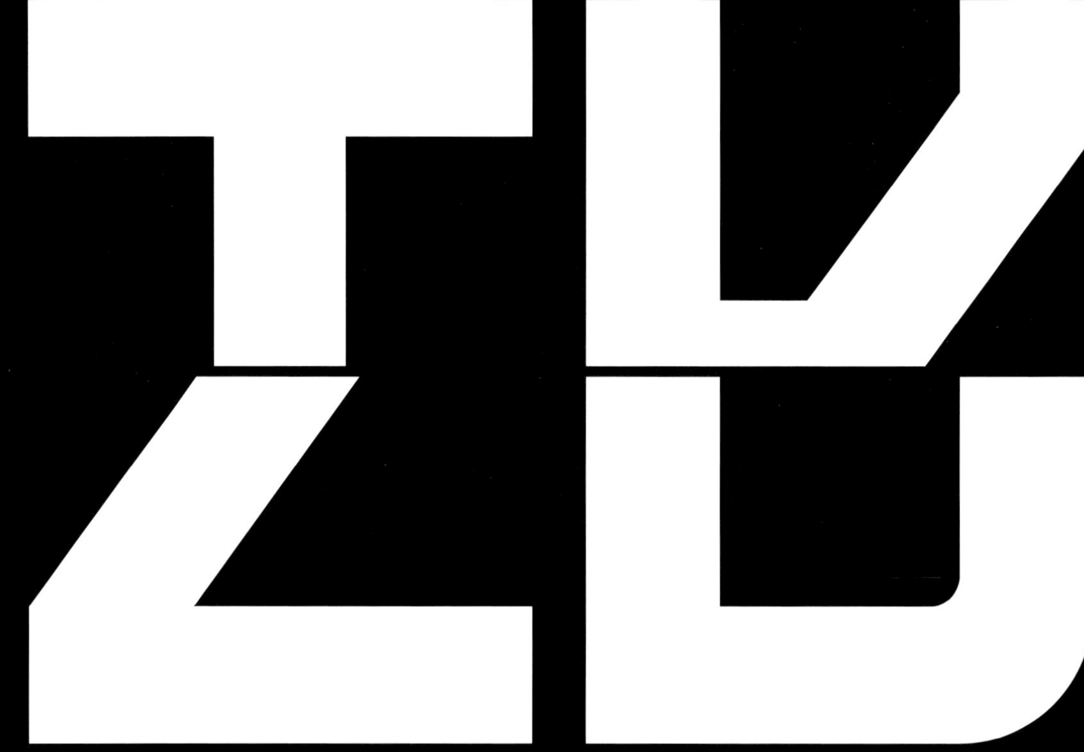

ABCDEFGHIJKLMNOPQRSTUVWXYZ

ACFAJLMPRTVXZ

0123456789

UDINE

Gestalter. Designer. David Dusanek
Jahr. Year. 2012
Webseite. Website. daviddusanek.de

Die systematische Rekombination inkongruenter Elemente setzt Interaktion und das Spiel mit Lesbarkeit in den Fokus des experimentellen Schriftentwurfs Udine. Alle lateinischen Buchstaben sowie alle arabischen Ziffern sind in einem Zeichensatz untereinander kombiniert, zu neuen Glyphen zusammengesetzt.

Der neu gezeichnete Monospace-Headlinefont mit seinem technischen Charakter zeichnet sich durch Stabilität und Naivität aus. In ihrer Grundform greift die Udine damit die Ästhetik historischer Industrieschriften, wie der britischen Kennzeichenschrift von Charles Wright auf. Initial entstand die Udine im Rahmen eines Kurses bei Prof. Johannes Bergerhausen (FH Mainz).

Über 1000 Glyphen sind über die Ligatur-Funktion nativ im Zeichensatz vorhanden, so dass sich das Schriftbild bereits während des Schreibens transformiert. Auch der akzentuierte Einsatz kombinierter Zeichen ist möglich und die Kontrolle über das Gleichgewicht zwischen Lesbarkeit und Irritation bleibt beim Gestalter. (dd)

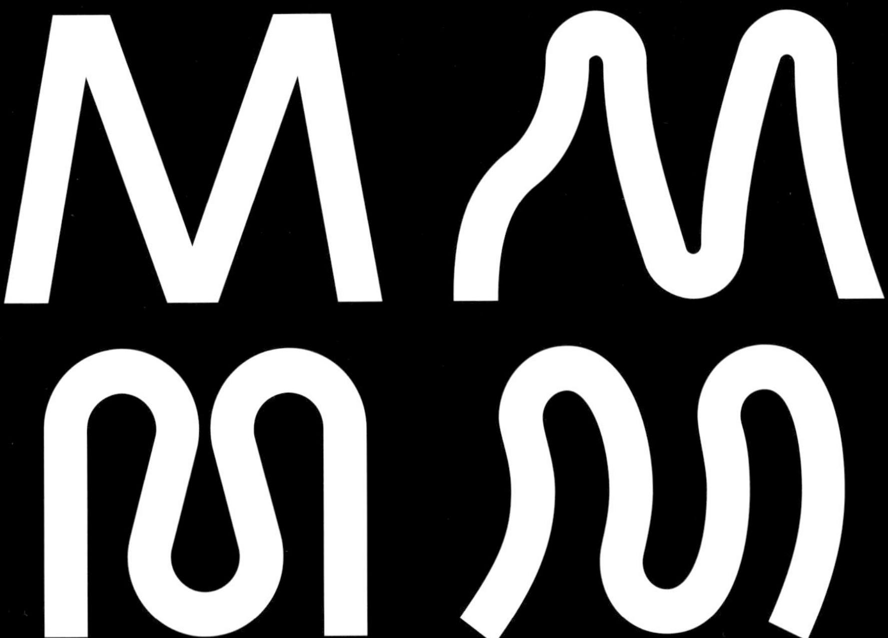

ABCDEFGHIJKLMNOPQRSTUVWXYZ
0123456789

ABCDEFGHIJKLMNOPQRSTUVWXYZ
ABCDEFGHIJKLMNOPQRSTUVWXYZ
ABCDEFGHIJKLMNOPQRSTUVWXYZ
0123456789

ZIGZAG

Gestalter. Designer. Benoît Bodhuin
Label. Foundry. Volcano Type
Jahr. Year. 2013
Webseite. Website. benbenworld.com

ZIGZAG Not Rounded by Benoît Bodhuin is a further development on ZIGZAG Rounded which was released in 2012.

ZIGZAG is a funny font family whose letters have four varieties each in order to multiply expressions and attract the eye by breaking the rhythm of reading.

The variations oscillate between a hand-drawn design and a geometric or imaginative drawing. OpenType functions let you choose between different variations of each glyph and contextual variables allow to mix the styles. (bb)

AVGVST

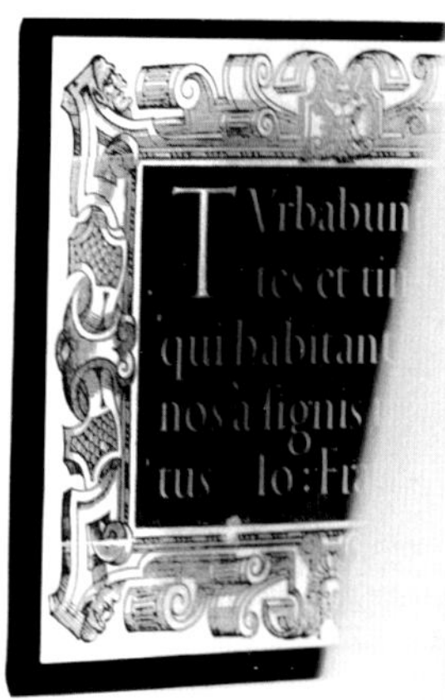
TVrbabun
tes et ti
qui habitan
nos a fignis
tus Io:Fr

VIA
DI S. GIOVANNI
IN LATERANO

ANHANG.
APPENDIX.

INDEX

123

A

B

C

D

E

F

G

H

I

J

K

DANK

DANK.
ACKNOWLEDGMENTS.

Allen Schriftgestalterinnen und -gestaltern, die sich an der Ausstellung und der Publikation beteiligt haben, sei herzlich gedankt. Durch die Leihgabe von Schriften und Begleitmaterialien sowie durch ihre Expertise und die intensiven Gespräche mit den studentischen Interviewpartnern haben sie das Projekt in äußerst großzügiger Weise unterstützt.

Der Jury des Wettbewerbs »Call for Type« gehörten an: Prof. Dr. Petra Eisele (Professorin für Designgeschichte und –theorie, Institut Designlabor Gutenberg / FH Mainz), Prof. Lars Harmsen (Professor für Konzeption / Typografie, FH Dortmund), Bernhard Hofmacher (Specialist Designer Liaison, Monotype), Akira Kobajashi (Type Director, Monotype), Dr. Annette Ludwig (Direktorin Gutenberg-Museum Mainz), Prof. Dr. Isabel Naegele (Professorin für Typografie und Gestaltungsgrundlagen, Institut Designlabor Gutenberg / FH Mainz) und Robin Scholz (Studentischer Vertreter FH Mainz).

Ein herzlicher Dank gilt Prof. Lars Harmsen, Bernhard Hofmacher und Akira Kobajashi, die gemeinsam mit uns mehr als 290 eingereichte Entwürfe aus aller Welt begutachteten, die nicht nur durch ihre Bandbreite, sondern auch durch ihre hohe Qualität beeindruckten. Ihre detaillierte Beurteilung und fachlich fundierte Bewertung verdient besondere Anerkennung.

Hoch motiviert und engagiert waren die Studierenden des Studiengangs Kommunikationsdesign der Fachhochschule Mainz an diesem Ausstellungs- und Publikationsprojekt beteiligt: Anna Alexander, Julia Bielefeld, Lyn Blees, Tabea Dölker, Matthias Dufner, David Dusanek, Jens-Peter Giesel, Lisa Grünwald, Bahar Hasan, Luzia Hein, Yvonne Kümmel, Felix Rank, Robin Scholz, Alexa Spors, Lisa Steinhauer und Tobias Villmeter. Sie haben sich sowohl gestalterisch als auch inhaltlich mit den »Neuen Schriften. New Typefaces.« auseinandergesetzt. Ihnen sind auch die Interviews in diesem Band zu verdanken.

Yvonne Kümmel und Jens-Peter Giesel zeichnen für die Gestaltung dieses Katalogs verantwortlich. Ihnen gilt unser besonderer Dank für ihren langen Atem und die gelungene Umsetzung. Die engagierte Ausstellungsgestaltung von »Call for Type« verdanken wir Simon Störk und Lukas Wezel. Das einprägsame Erscheinungsbild der Ausstellung stammt von Matthias Dufner und David Dusanek. Unser Dank schließt darüber hinaus die Mitarbeiterinnen und Mitarbeiter des Gutenberg-Museums sowie alle am Projekt Beteiligten mit ein. Kerstin Forster, Verlag Niggli, danken wir für die vertrauensvolle Zusammenarbeit.

Für finanzielle Unterstützung sind wir der Landeshauptstadt Mainz, dem Förderverein Gutenberg e.V., der Fachhochschule Mainz und ihrem Präsidenten sowie den nachfolgend aufgeführten Sponsoren zu Dank verpflichtet: Bengsch & Störk Projektmanagement, Botschaft des Königreichs der Niederlande, designkritik dk, descom, Monotype, Selekkt.com, Slanted.

Die Herausgeberinnen

Our thanks go to all type designers participating in this exhibition and publication, who so generously supported our project through their loan of typefaces and associated materials, but also by their expertise and readiness to grant in-depth interviews to our student interview partners.

Members of the Jury for »Call for Type« were Prof. Dr. Petra Eisele (Professor for Design History and Theory, Institut Designlabor Gutenberg, FH Mainz), Prof. Lars Harmsen (Professor for Conception / Typography, FH Dortmund), Bernhard Hofmacher, (Specialist Designer Liaison, Monotype), Akira Kobajashi (Type Director, Monotype), Dr. Annette Ludwig (Director Gutenberg-Museum, Mainz), Prof. Dr. Isabel Naegele (Professor for Typography and Design Principles, Institut Designlabor Gutenberg / FH Mainz) and Robin Scholz (Student representative, FH Mainz).

More than 290 design samples were submitted from all over the world, covering a wide range of styles and convincing through their high quality. Therefore, our particular thanks go to Prof. Lars Harmsen, Bernhard Hofmacher, Akira Kobajashi who, in cooperation with the three project leaders, gave their carefully detailed evaluation and well-founded judgement.

Our thanks also go to the following students of the programme for Communication Design at the University of Applied Science, Mainz, who carried the project by their high motivation and deep involvement: Anna Alexander, Julia Bielefeld, Lyn Blees, Tabea Dölker, Matthias Dufner, David Dusanek, Jens-Peter Giesel, Lisa Grünwald, Bahar Hasan, Luzia Hein, Yvonne Kümmel, Felix Rank, Robin Scholz, Alexa Spors, Lisa Steinhauer and Tobias Villmeter, who were grappling with »Neue Schriften. New Typefaces.« both from the design and content point of view. We also thank them for the interviews in this publication.

Yvonne Kümmel and Jens-Peter Giesel were responsible for the design of this catalogue, and we would like to thank them for their patience and successful execution. The exhibition »Call for Type« was staged by Simon Störk and Lukas Wezel, and the eye-catching Corporate Design was developed by Matthias Dufner and David Dusanek. Our gratitude includes all parties involved, in particular also the staff of the Gutenberg-Museum. Special thanks go to Kerstin Forster / Niggli for her support and confidence.

Finally, our thanks go to the City of Mainz, the Förderverein Gutenberg e.V. , the University of Applied Sciences, Mainz and its President for their financial support, as well the following sponsors: Bengsch & Störk Prokektmanagement, the Embassy of the Netherlands, designkritik dk, descom, Monotype, Selekkt.com, Slanted.

The Editors

IMPRESSUM

AUSSTELLUNG.
EXHIBITION.

Eine Ausstellung des Gutenberg-Museums Mainz und des Instituts Designlabor Gutenberg der Fachhochschule Mainz im Gutenberg-Museum, 7. Juni – 27. Oktober 2013

www.call-for-type.de

Museumsdirektorin. Director Museum.
Dr. Annette Ludwig
Liebfrauenplatz 5, D-55116 Mainz
www.gutenberg-museum.de

Konzeption & Projektleitung.
Concept & Project Direction.
Prof. Dr. Isabel Naegele, Prof. Dr. Petra Eisele, Dr. Annette Ludwig

Assistenz & Pressearbeit.
Assistance & Press Relations.
Robin Scholz, Dr. Juliane Schwoch

Ausstellungsgestaltung.
Exhibition Design.
Simon Störk, Lukas Wezel

Kommunikationsdesign.
Communication Design.
Matthias Dufner, David Dusanek

Programmierung. Programming.
Matthias Dufner, Christian Hansen

Mit freundlicher Unterstützung von.
With kind support from.
Förderverein Gutenberg e.V.,
Bengsch & Störk Projektmanagement GmbH,
Monotype, Königreich Niederlande, Slanted,
Designkritik dk, descom

PUBLIKATION.
PUBLICATION.

Neue Schriften. New Typefaces.
Erscheint als Dokumentation der Ausstellung
»Call for Type. New Typefaces. Neue Schriften.«
Gutenberg-Museum Mainz
7. Juni – 27. Oktober 2013

Herausgegeben von. Edited by.
Gutenberg-Museum Mainz, Dr. Annette Ludwig
Institut Designlabor Gutenberg / Fachhochschule Mainz,
Prof. Dr. Isabel Naegele, Prof. Dr. Petra Eisele

Konzeption & Redaktion.
Concept & Editing.
Isabel Naegele, Petra Eisele, Annette Ludwig

Interviews. Interviews.
Studierende des Studiengangs Kommunikationsdesign
der Fachhochschule Mainz:
Anna Alexander, Julia Bielefeld, Lyn Blees, Tabea Dölker,
Matthias Dufner, David Dusanek, Jens-Peter Giesel,
Lisa Grünwald, Bahar Hasan, Luzia Hein, Julia Heil,
Yvonne Kümmel, Felix Rank, Robin Scholz, Alexa Spors,
Lisa Steinhauer, Tobias Villmeter

Übersetzung. Translation.
Hannelore Müller, Christine Naegele

Gestaltung. Graphic Design.
Jens-Peter Giesel, Yvonne Kümmel

Fotografie. Photography.
Lars Harmsen, Joseph Kadow, Michael Schmitz

Schriften. Typefaces.
Suisse B+P (Ian Party)

Papier. Paper.
Multi Art Gloss

Druck. Print.
Heer Druck AG, Sulgen

Buchbinder. Binding.
Kösel GmbH, Altusried-Krugzell

Bibliografische Informationen der Deutschen Nationalbibliothek. Die Deutsche Nationalbibliothek verzeichnet diese Publikation in der Deutschen Nationalbibliografie; detaillierte Informationen sind im Internet über http://dnb.d-nb.de abrufbar.

Bibliographical information published by the Deutsche Nationalbibliothek. The Deutsche Nationalbibliothek lists his publication in the Deutsche Nationalbibliografie. Detailed bibliographical information is available at http://dnb.d-nb.de.

Erschienen bei. Published by.

niggli Verlag
Sulgen
www.niggli.ch

ISBN 978-3-7212-0892-4

IMPRINT